The Patrick Moore Practical Astronomy Series

W0227506

For further volumes:
http://www.springer.com/series/3192

Blazing a Ghostly Trail

ISON and Great Comets of the Past and Future

Peter Grego

 Springer

Peter Grego
St Dennis, Cornwall, UK

ISSN 1431-9756
ISBN 978-3-319-01774-7   ISBN 978-3-319-01775-4 (eBook)
DOI 10.1007/978-3-319-01775-4
Springer Cham Heidelberg New York Dordrecht London

Library of Congress Control Number: 2013950100

Printed on acid-free paper

Springer is part of Springer Science+Business Media (www.springer.com)

Preface

Right now, a comet is heading towards the inner Solar System. This visitor from the frozen cosmic depths—an 'icy dirtball' the diameter of a small town—has every chance of becoming bright enough to be seen with the unaided eye as the increasing heat of the Sun turns its ices to gas and its dust grains are freed into space. Indeed, the comet is likely to develop spectacular gas and dust tails that stretch across the sky. One of nature's most sublime sights, its appearance is set to be indelibly impressed upon the long-term memories of everyone who sees it.

This was by no means an exaggerated statement. It has always applied, since before the first human set eyes upon a comet. There's little chance that the supply of comets from the far-frozen depths of space will ever dry up. There's always a comet heading our way, and really bright ones dazzle us by surprise every few decades or so.

Between June 1872 and April 1910 Halley's Comet (1P/Halley) was on the inbound leg of its 75-year orbit, having passed its furthest point from the Sun (aphelion). Astronomers knew enough then to be able to predict the comet's path with a good degree of accuracy and were able to confidently claim—based on the past performance of Halley's Comet—that it would present quite a stunning sight in the skies. But how can astronomers predict that any newly discovered comet is likely to dazzle?

Our opening paragraph applies to my main prompt for writing this book—the discovery of a potentially spectacular comet by the official designation C/2012 S1 (ISON), which will be introduced presently.

First, though, a few words of caution. Having had a deep fascination with the beauty of the night skies since I can remember, my introduction to "space" was doubtless much like that experienced by many readers of this book—initially arising from a notable celestial event that was publicized in print, on television or radio. Comets are beloved by the media, and they often hit the headlines because they are sometimes predicted to grow into spectacular sights. You may be old enough to have had your interest piqued by the media hype surrounding Comet Kohoutek (C/1973 E1). Your interest in astronomy may have snowballed from a far more recent event, say the multiple impacts on the giant planet Jupiter of fragments of Comet Shoemaker-Levy 9 (D/1993 F2) in July 1994, or by the magnificent showing of Comet Hale-Bopp (C/1995 O1) in 1997. Perhaps your curiosity about astronomy resulted from an altogether different space event, say, watching the late Neil Armstrong plant his size 11 Moon boot on the Sea of Tranquility in July 1969, or just by having seen pictures from the Hubble Space Telescope. Maybe this is the first astronomy book you've ever read, in which case I hope to do a good job in kindling the flames of cosmic and cometic enquiry within you.

It's my intention that this book will stand as a handy guide to comets in general, long after C/2012 S1 has come and gone. We'll take a look at what astronomers have learned about comets throughout history, discovering what comets are and where they come from. Our journey will take us on a whistle-stop tour of Great Comets and remarkable comets of the past, some recent comets and the prospects of comets to come. And we'll go out together under the dark, starry skies, to talk about how you can enjoy observing comets, whether you're a casual stargazer or serious amateur astronomer with a desire to "capture" a comet for yourself.

Since May 1983 I have observed and sketched 27 different comets, all of which have been bright enough to be seen through binoculars. Nevertheless, 27 comets in 30 years isn't a particularly impressive tally to the truly serious comet observer; in reality, the number of comets that could have been glimpsed in this period from my location in the UK with the instruments that I had available is perhaps double this number, but most of these extra ones would have appeared pretty faint.

These experiences in stargazing have stood me in good stead, not only in knowing what to reasonably expect to see through the eyepiece but also to expect the unexpected. Comets are often fickle, and they have absolutely no knowledge of how bright astronomers have predicted they will become. Some exceed all expectations by outshining brightness predictions and developing detailed structure in their heads and tails. Some behave in remarkable ways by flaring suddenly, developing anti-tails, by breaking up or even dashing themselves to smithereens on planets. Those who remember the hyperbole surrounding Comet Kohoutek that was promulgated in the media also know what disappointment means. This has tempered my approach in writing this book.

Stand By for Action

As an astronomical editor and writer I try to keep abreast of what's happening in space; neglecting to flag-up a particularly rare or spectacular astronomical event, such as a prospective naked-eye comet, would do little to enamor myself to my readers.

On September 24, 2012, an electronic circular issued by the Minor Planet Center (MPC) flashed up in my inbox. *MPC Electronic Circulars* (*MPECs*) detail discoveries and observations of small Solar System objects, such as comets, asteroids and errant satellites. They are, however, issued frequently—at least once a day—and, quite frankly, most are of little relevance to my target readership.

MPEC 2012-S3 was, like all the others, chiefly a list detailing an object's discoverers/observers, along with various data and orbital parameters for that object. A scientific document, it contained no hyperbole; there was no friendly editorial commentary, and it featured no charts or images. Just the facts. *MPEC 2012-S3* detailed a newly discovered comet, designated C/2012 S1 (ISON), which had first been spotted on September 21 by astronomers Vitali Nevski and Artyom Novichonok using the 0.4-m (16-in.) reflector of the International Scientific Optical Network (ISON) near Kislovodsk, Russia. Images taken from other observatories prior to and after the discovery had enabled its orbit to be determined with some accuracy.

It soon became clear to me that the numbers suggested something extraordinary about this comet, and after a little more investigation the hair on the back of my neck began to stand up. Doubtless thousands of the *MPEC's* recipients—even the most impassionate of astronomers—experienced a similar visceral thrill on realizing how special this newly discovered comet was. Incredibly far away on discovery, C/2012 S1 was already bright enough to be imaged through relatively modest amateur equipment. Pulled by the Sun's gravity it was coming in from deep space— probably on its first visit to the inner Solar System since its formation 4.6 billion years ago. Brightening as it plunged into the heart of the Solar System, it was predicted to swing very closely around the Sun in November 2013. In September 2012 I reported:

> A new comet has been discovered that is predicted to blaze incredibly brilliantly in the skies during late 2013. With a perihelion passage of less than two million kilometres from the Sun on 28 November 2013, current predictions are of an object that will dazzle the eye at up to magnitude −16. That's far brighter than the full Moon. If predictions hold true then C/2012 S1 will certainly be one of the greatest comets in human history, far outshining the memorable Comet Hale-Bopp of 1997 and very likely to outdo the long-awaited Comet Pan-STARRS (C/2011 L4) which is set to stun in March 2013. Its near-parabolic orbit suggests that it has arrived fresh from the Oort Cloud, a vast zone of icy objects orbiting the Sun, pristine remnants of the formation of the Solar System.

> C/2012 S1 currently resides in the northwestern corner of Cancer. At magnitude +18 it is too dim to be seen visually, but it will be within the reach of experienced amateur astronomers with CCD equipment in the coming months as it brightens. It is expected to reach binocular visibility by late summer 2013 and a naked eye object in early November of that year. Northern hemisphere observers are highly favoured. Following its peak brightness in late November it will remain visible without optical aid until mid-January 2014.

> Comet brightness predictions sometimes exceed their performance. Amateur astronomers of a certain age may remember the Comet Kohoutek hype of 1973 – not quite the 'damp squib' it has been portrayed, since it reached naked eye visibility! Even if C/2012 S1 takes on the same light curve as Kohoutek it is certain to be spectacular, quite possibly a once-in-a-civilisation's-lifetime event.

Now writing in June 2013, I stand by the above, even though it may to some appear to wander perilously near to hyperbole. But there's no harm in being given ample warning of a coming astronomical event, and "Be Prepared" is as good a motto for the amateur astronomer as it is the Boy Scout. Some amateur astronomers had prepared observing plans for Halley's Comet several years before it became visible through the telescope in 1985; indeed, the International Halley Watch was endorsed by the International Astronomical Union (IAU) as far back as 1982.

As we shall see, newly discovered comets—particularly those on their first visit to the inner Solar System—can behave unpredictably, and astronomers are at pains to offer the caveat that brightness predictions rely upon a range of assumptions. Some new comets, such as Comet Kohoutek, ticked all the right boxes for becoming a dazzling object, but failed to live up to optimistic predictions; 3 years later, however, Comet West exceeded all expectations and became a brilliant object in February 1976. Sure, some comets can be fickle, and we'll be taking a look at some of the reasons why.

In *Julius Caesar*, Shakespeare wrote:

When beggars die, there are no comets seen;
The heavens themselves blaze forth the death of princes.

In the same play, Shakespeare gave these lines to his eponymous hero:

But I am constant as the northern star,
Of whose true-fixed and resting quality
There is no fellow in the firmament.

I humbly suggest a few lines from my own (as yet unpublished and probably never to be published) play *Caligula*, after the style of the immortal Bard:

Inconstant as the comet's luster,
My mind like Luna waxes-wanes.
Let my reason recede that my passions flare!
Let them hate me, so that they will but fear me
As, hairs on end they fear the comet yet do its hairy tail admire.
(apologies to all true lovers of the Bard of Avon)

St Dennis, Cornwall, UK   Peter Grego

Acknowledgments

I could not have completed this book without an immense amount of support and encouragement from Maury Solomon, Senior Editor of Professional and Practical Astronomy, Astrophysics, Astronautics, and Space Exploration at Springer in New York. Maury's patience with me extended far beyond the call of duty. Although logic kept on telling me that Maury would pull the plug after each delivery promise I made was unfulfilled, she held out and things came good in the end.

It is to Maury that I dedicate this book.

Thanks are also due to John Watson, Nora Rawn, and the team at Springer.

My friend and long-time occasional observing pal Paul Stephens has been a help and an inspiration. Together we have admired many a comet during the course of the past three decades.

Contents

Chapter 1

What Are Comets and Asteroids

Let's begin by describing comets as we currently understand them. In popular culture, as seems to have always been the case, comets are simply bad news. Modern entertainment media (and, sadly, some sensationalist news media) are fond of misrepresenting them. Cocooned in a growing nebula of gas and dust, their trailing tails growing longer night by night and they hang in the sky like the sword of Damocles. They possess an almost conscious intention to hit something—and that something is usually Planet Earth (Fig. 1.1).

Here are some "facts" about comets as portrayed in films and the popular media:

- *They are big*. They range in size from a mountain to a moon.
- *They are solid*. They are made of metals, rock, ice, or a combination of the three.
- *They hurtle through space*. They are going so fast that they can catch humanity unawares and unable to patch a plan together to stop them if need be.
- *They hit hard*. They pack a punch hard enough to take out anything from a city block to an entire continent.
- *They are bad news*. At best they can cause general alarm and panic; at worst they can produce an extinction level event.

That's entertainment, folks! But in exploring the facts, we'll find plenty of entertainment—more, perhaps, than any fiction writer could conjure.

P. Grego, *Blazing a Ghostly Trail: ISON and Great Comets of the Past and Future*,
The Patrick Moore Practical Astronomy Series, DOI 10.1007/978-3-319-01775-4_1,
© Springer International Publishing Switzerland 2014

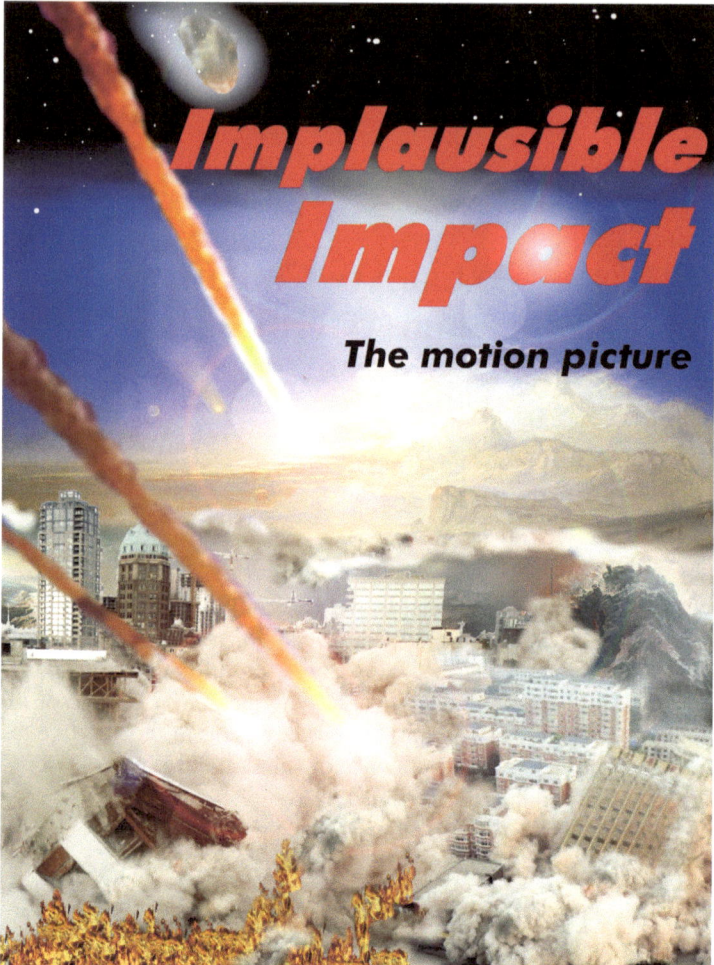

Fig. 1.1 Fear and panic—a good way of selling a movie about comets (Illustration © by the author)

Icy Dirtballs

Comets belong to a group of objects classified as small Solar System bodies (SSSB), a group that contains—with the exception of the eight planets, dwarf planets and planetary satellites—every solid object in the Solar System orbiting the Sun. SSSBs include comets, all the minor planets (asteroids), meteoroids and, technically, micrometeoroids (interplanetary dust).

Cometary nuclei—the solid objects that lie at the heart of all comets—range from several hundred meters to several tens of kilometers across; they are irregularly

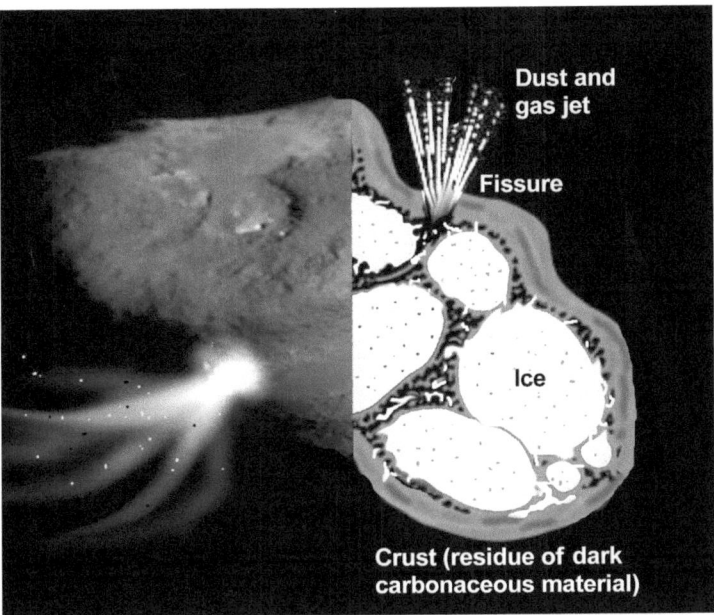

Fig. 1.2 Cross-section through a typical active cometary nucleus (Illustration © by the author)

shaped, since their low mass prevents them from gravitationally forming a spherical shape. Among the darkest objects in the Solar System, nuclei typically have an albedo (reflectivity) of just 0.04; reflecting just 4 % of the sunlight falling on them; they can appear as dark as a fresh asphalt surface. As a result of their small size and low surface brightness, nuclei are exceedingly faint and difficult to observe through Earth-based telescopes, especially when they are much further away than the orbit of Neptune, the most distant planet from the Sun at 30 au. The au, or astronomical unit, is a convenient scale upon which astronomers measure distances in the Solar System. One au is equivalent to the average distance between the Sun and Earth, some 149,597,871 km (Fig. 1.2).

Cometary nuclei are solid agglomerations of silicate material (the sort of stuff that makes up Earth's solid surface) in the form of dust and small rocky particles, and ices—mainly water ice, but also various proportions of frozen ammonia, methane, nitrogen, carbon monoxide and carbon dioxide. A variety of organic (carbon-based) compounds can also be found in the mix, including ethane, ethanol, formaldehyde, hydrogen cyanide and methanol; indeed, it is likely (and has long been speculated) that more complex organic molecules, such as amino acids, form part of some comets' makeup (Figs. 1.3 and 1.4).

Somewhat ignominiously, comets have been described as 'dirty snowballs'; a more appropriate description might be 'icy dirtballs.' It's a subtle difference, but in general it's an important one. An active, long-lived cometary nucleus can't be fluffy

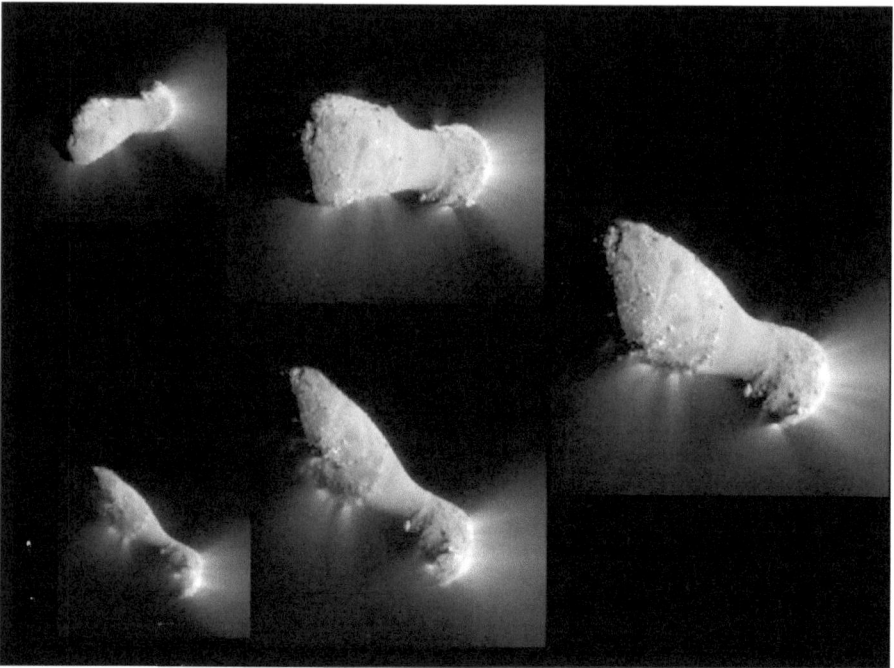

Fig. 1.3 Note the incredible activity in the nucleus of Comet Hartley 2 (103P/Hartley), imaged by NASA's Deep Impact flyby (EPOXI) (NASA)

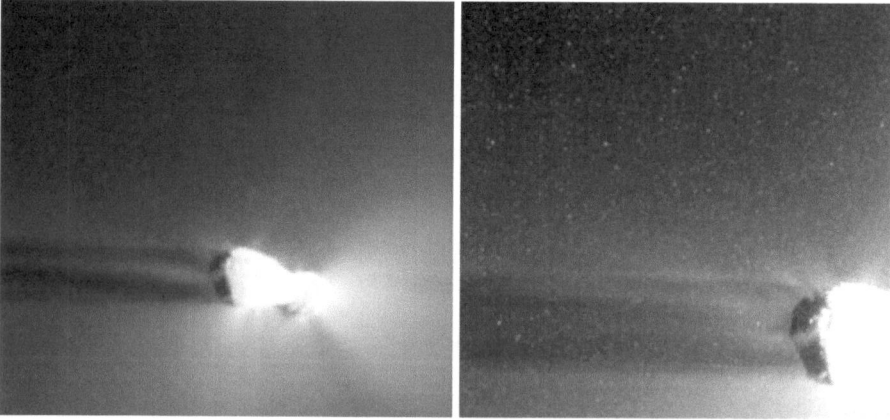

Fig. 1.4 An enhanced view of Comet Hartley 2 (103P/Hartley), imaged by NASA's Deep Impact flyby (EPOXI), showing the shadow cast by its nucleus (NASA)

and insubstantial like a playfully patted kindergarten snowball; such objects might well exist far out in the extreme chilly depths of space, even quite small ones, lightly held together as a single mass by their own gravity. But any such snowball compelled to venture from its distant lair and into the inner Solar System would be tugged apart by the gravity of the Sun and planets and quickly melted away into nothing by the Sun's heat. In order that it holds together as a solid mass during its travels through the Solar System, particularly if it hangs around the planetary realms for any length of time and experiences multiple orbits around the Sun, a cometary nucleus must be robust enough to withstand gravitational tugging by the Sun and planets.

Comets are thought to have been formed out of the original interstellar material that condensed to form the Sun and the Solar System, although it is possible that a few may have originated from nearby star systems. Comets therefore provide essential clues to the composition of the original cloud from which the Solar System evolved.

Comet Comas and Tails

If a cometary nucleus ventures into the inner reaches of the Solar System, it begins to respond to increased levels of solar heat and energy. Heated by the Sun, the mixture of ices on the surface of the nucleus change from their solid state to a vapor (gas). The pressure is far too low for the ices to go through an intermediate liquid phase. At around the distance of Saturn, some 10 au from the Sun, the heat received by the nucleus is sufficient to cause the most volatile ices on the surface—carbon monoxide and carbon dioxide—to sublimate. Water ice begins to sublimate at a distance of around 3 au, around the distance of the main Asteroid Belt between Mars and Jupiter (Fig. 1.5).

The tiny bright spot seen at the center of the coma in views from Earth is a known as a false nucleus, its brightness being produced by the concentration of gases and dust emanating from the nucleus' surface. Although some comets (especially those at some distance from the Sun) may appear as nebulous patches, with little obvious structure besides a certain brightening towards an offset brighter central area, most comets display a false nucleus. Through the telescope and on images taken from Earth, the false nucleus may appear as a faint, star-like point in dimmer comets, while a large, bright, active comet may have a false nucleus with an apparent angular diameter of several arcseconds.

Gases released from the surface of the comet's nucleus are bombarded by the solar wind, a 500 km/s stream of electrons and protons that is continually expanding away from the Sun into the depths of interplanetary space. The comet's gases ionize as they pick up electrical charge from the solar wind, and these ionized gases form a large glowing cloud, known as a coma, around the nucleus. The coma can grow to several hundreds of thousands of kilometers in diameter. Composed of charged particles (protons and electrons), the solar wind carries with it a magnetic field, and

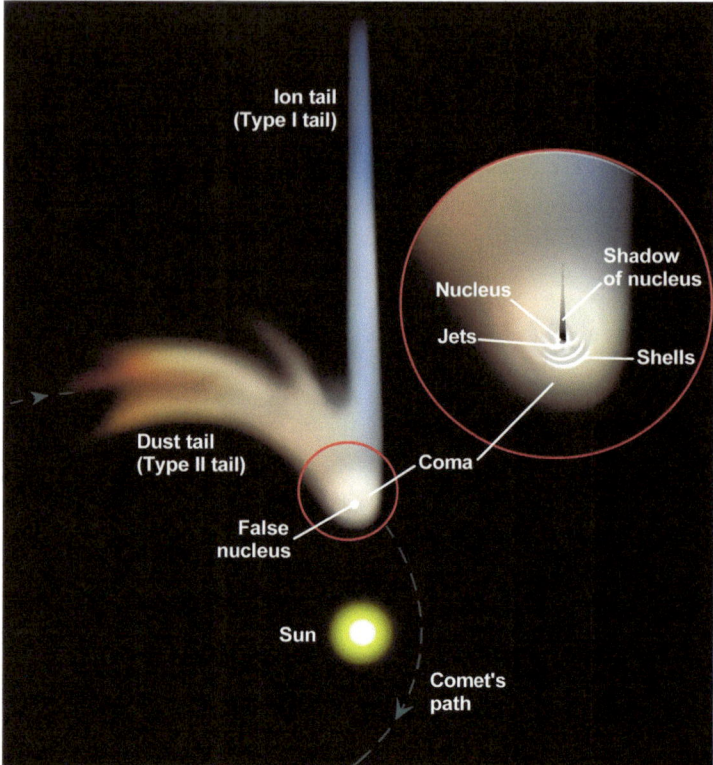

Fig. 1.5 The main components of a comet and a close-up of the coma and nucleus. The nucleus itself and the shadow it casts can't be seen using Earth-based instruments (Illustration © by the author)

the resistance produced by the coma's electrified gases presents a barrier to the lines of magnetism. As a result, the magnetic field lines are deflected around the comet to drape around the comet's head, forcing the gases to stream directly away from the Sun. The ion tail, known as a Type I tail, is generally straight and points directly away from the Sun, regardless of the actual motion of the comet through space; seemingly counter-intuitively, a comet races in the general direction of its ion tail following perihelion. Type I tails can take on a distinctly blue color—a hue largely caused by the fluorescence of ionized carbon monoxide as it is excited by the Sun's ultraviolet light—and they make for a striking sight through binoculars and on images.

Type I tails are susceptible to intricate interaction with charged particles in the solar wind, and they often display remarkable twists and kinks as they are buffeted by more vigorous gusts. On occasion the solar wind can produce a blast so powerful that the magnetic field lines flowing around the comet are forced to reconnect with themselves on the far side of the nucleus, producing a brief burst of energy

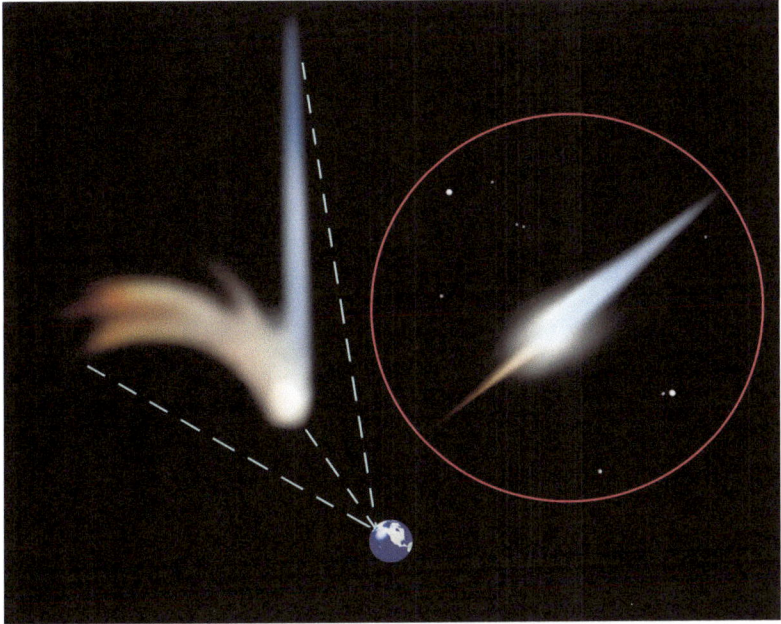

Fig. 1.6 The anti-tail (inset, a view from Earth) is a line-of-sight illusion caused when a comet is viewed from an angle where the curving end of a Type II tail appears to point in the opposite direction to the comet's main tail (Illustration © by the author)

sufficient to disconnect the ion tail completely. Of course, disconnection events are only transient occurrences. As the glow of the old ion tail fades and drifts away with the solar wind, the comet develops another ion tail in its place.

Rather than sublimating evenly across its sunlit face, the comet's nucleus sublimates most vigorously in certain locations around the nucleus, giving rise to jets, streams of gas that bring with them dust grains and rocky particles that were originally embedded in the icy crust. Pressure from the solar wind causes the released dust grains to lag behind the nucleus, producing a separate dust tail—known as a Type II tail—that develops alongside the gaseous ion tail. Composed of a myriad of dust particles and rocky fragments released from the nucleus and shining by reflected sunlight, the Type II dust tail of a bright comet often assumes a yellow or orange-brown hue, complementing the blue color of the ion tail.

Each dust particle released by the nucleus continues along the same general path as the comet, but solar radiation pressure pushes the particles away from the Sun. Forced to lag behind, the particles create a tail that appears distinctly curved. Anti-tails—apparent 'spikes' that appear to protrude from some comets in a direction opposite the main tail—are simply line-of-sight effects, produced when Earth is level with the comet's orbital plane and distant parts of the curving dust tail appear to precede the comet's head (Fig. 1.6).

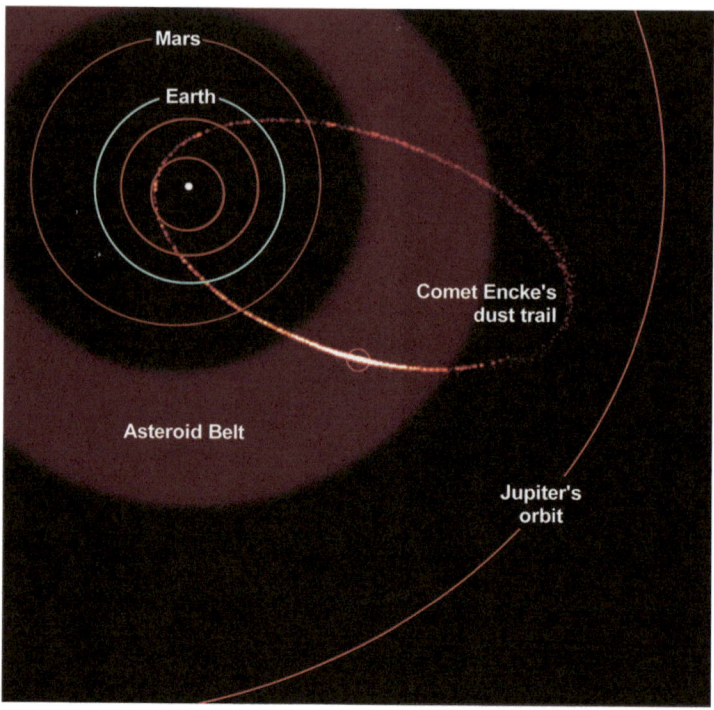

Fig. 1.7 A simulated view of the dust trail left by Encke's Comet. November's Taurid meteors are produced when Earth intersects the stream (Illustration © by the author)

The dust particles released by comets provide the stuff of meteoroid streams and the annual meteor showers visible at certain times throughout the year (described later). In 1983 the satellite IRAS (Infrared Astronomical Satellite) actually detected more than a hundred faint trails of meteoroidal dust within the Solar System, and identified distinct trails associated with the comets Encke, Tempel 2 and Gunn. Comet Encke's dust trail (from which is derived November's Taurid meteor shower) was measured on IRAS images to be 300 million km long—equal to the diameter of Earth's orbit around the Sun (Fig. 1.7).

Most comets develop both a Type I ion tail and a Type II dust tail, although their real and apparent size and form varies enormously with each comet and the angle at which it is viewed. Comet tails can grow to enormous lengths of millions, tens of millions or (rarely) hundreds of millions of kilometers. In 1997, the remarkable Comet Hale-Bopp developed a strikingly blue Type I tail and a curving, orange Type II tail. In addition, astronomers using specialist filters were startled to discover, for the first time, a very faint third kind of tail—one composed of sodium atoms—streaming between the ion and dust tails. Unlike the atoms of the Type I tail, which form an ionized gas, the sodium atoms of this new type of tail—now known as a Type III tail—are neutral atoms whose origin remains to be satisfactorily explained; whatever their source, they are driven away from the nucleus to form a tail by the pressure of radiation from the Sun.

As the comet approaches perihelion (its closest point to the Sun), the most frenetic phase of the process begins. In large active comets the false nucleus is often noticeably asymmetric in shape and may extend into bright streamers. These features are produced by active jets on the surface of the nucleus; as the nucleus rotates, the streamers assume a curved or swirling pattern. At the same time, areas of potential activity are brought into the sunlight, producing variations in the brightness and shape of the false nucleus over a period of a few hours. Material ejected by an active area flows away from the nucleus, and as its supply zone is cut off when it rotates onto the inactive dark side of the nucleus, a shell of material may form on the sunward side of the nucleus. Several concentric layers of shells may be produced, each curving away from the Sun with the bow shock of the solar wind, as fresh ejections of material are sprayed out from the nucleus.

Comets vary enormously in the amount of dust and gas that they produce. Some comets are dustier than others, while some old comets, veterans of many perihelia, have built up substantial crusts of silicate material that prevent much of their internal icy content from being heated by the Sun and sublimated. Although some comets display a simple, broad, fuzzy globular coma, others may go on to develop a magnificently structured coma, plus lengthy, bright ion and dust tails. Some of the more active comets become really bright and spectacular objects, and if conditions are right they can be truly impressive celestial spectacles for a period of time. Sadly, the majority of comets discovered never reach naked eye visibility and remain dim objects visible only with optical aid.

Comet Orbits

As Kepler discovered four centuries ago, all planetary orbits are ellipses of varying eccentricity, with the Sun located at one focus of this ellipse. If drawn to scale on a piece of paper you might find it difficult to appreciate that Earth is in anything but a circular orbit around the Sun. That's because its eccentricity is just 0.0167 (0 being a perfect circle). Even the most eccentric of planets, Mercury (eccentricity 0.2056), has an orbit that, at first sight, looks circular when sketched out on paper.

Comets also obey Kepler's first law of planetary motion, but most comets have highly elliptical orbits with eccentricities approaching 1—far greater than any of the planets—and, furthermore, their eccentricity increases with increased orbital period. For example, Halley's Comet has an orbital eccentricity of 0.967 and a period of 75.3 years, while Comet Hale-Bopp has an eccentricity of 0.995 and a period of 2,500 years. Halley's Comet is the best-known example of a "short-period" comet—one with an orbital period of less than 200 years. Hale-Bopp is a particularly splendid example of a "long-period" comet, a term that applies to comets with periods longer than 200 years (often many thousands, even millions of years) that are gravitationally bound to the Sun. Much rarer, "single-apparition"

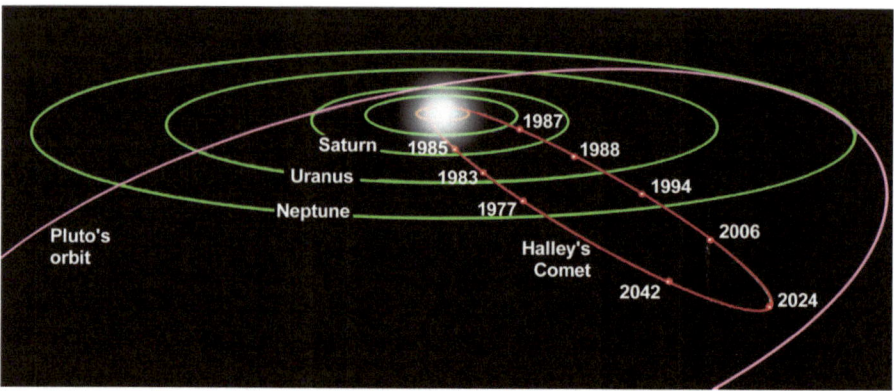

Fig. 1.8 The entire orbit of Halley's Comet is within our planetary realm (Illustration © by the author)

comets are those with near-perihelion orbital eccentricities exceeding 1—that is, their orbits are hyperbolic (a curved path that fails to close into a complete figure)—which means that they can escape the Solar System altogether following perihelion (Fig. 1.8).

A Distant Cometary Realm

Newly discovered comets can fall into the Solar System from virtually any direction, but there's a marked tendency for long-period comet orbits to originate at a distance of about 50,000 au from the Sun. Their origin, far beyond the planets—in realms so distant that the Sun appears as a bright star-like point—is a zone populated by countless icy cometary nuclei. Marking the very outer fringes of the Sun's gravitational domain, this zone is known as the Oort Cloud, named after Jan Oort (1900–1992), a Dutch astronomer who first postulated its existence in the midtwentieth century (Fig. 1.9).

Individual cometary nuclei lying deep within the Oort Cloud have never been directly observed, nor do we currently have much prospect of detecting them at such enormous distances. Although the precise size and shape of the Oort Cloud remains unknown, astronomers have gained some idea of its dimensions from plotting the orbits of occasional cometary visitors from the depths of space. Its inner region is speculated be a flattened disc, broadly aligned with the ecliptic plane extending from around 50,000 au, which broadens into a spherical outer zone that is elongated towards the galactic center, measuring around $200,000 \times 160,000$ au—a vast shell that extends around a quarter of the way to the nearest stars. The combined mass of comets within the Oort Cloud is estimated to

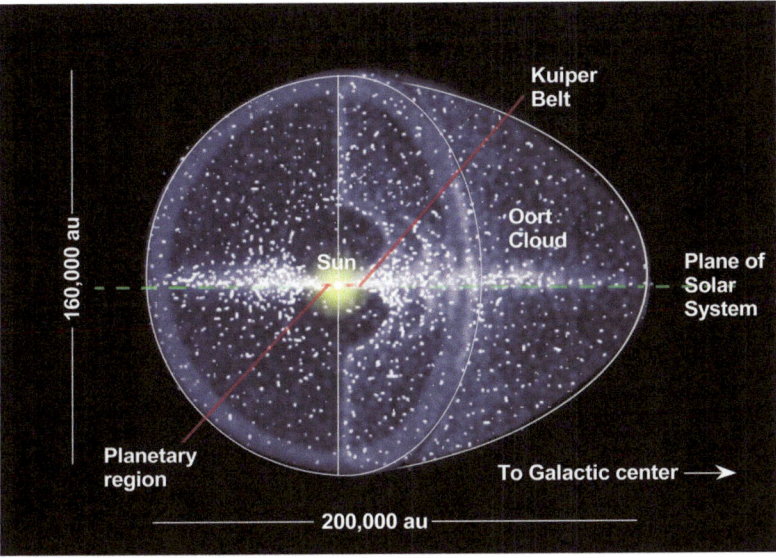

Fig. 1.9 A cross-section through the Oort Cloud, a vast, distant region populated by cometary nuclei (Illustration © by the author)

be as high as 50 times the mass of the Earth. That's a lot of material, making up several trillions of cometary nuclei. Most Oort Cloud comets lie within its denser inner zone.

Comets within the Oort Cloud are, on average, many tens of millions of kilometers apart. Despite the enormous numbers of comets there, the sheer volume of the Oort Cloud means that it is far more sparsely populated than the main Asteroid Belt between Mars and Jupiter. In the Oort Cloud's outer realms, comets are so weakly held by the Sun's gravity that their orbits can be perturbed by the gravity of nearby stars. As a result, comets may be nudged into a different orbit within the Oort Cloud, flung off into interstellar space or dispatched into the inner Solar System. In the latter case, we finally get to view them as they swing through the inner Solar System and switch on.

It is thought that every half billion years or so the Sun, carrying the Solar System and the Oort Cloud along with it, cuts a swathe through vast accumulations of interstellar hydrogen gas in the galactic arms. The gravitational attraction of these giant molecular clouds distorts the weakly bound outer shell of the Oort Cloud, causing a significant disruption in the orbits of the outer Oort comets. Significant influxes of comets into the inner Solar System may result on a cyclical basis from such events, increasing the likelihood of collisions with planets. Indeed, cometary impacts have been cited as one possible cause of the periodic mass extinctions in Earth's history (Fig. 1.10).

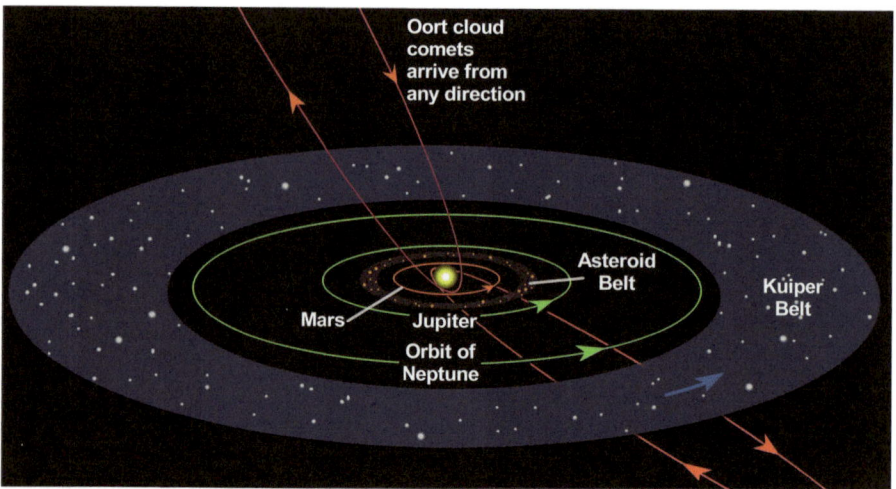

Fig. 1.10 Oort Cloud comets can arrive from any direction, while the planets, main belt asteroids and most Kuiper Belt comets orbit the Sun anticlockwise (viewed from north) and in the same general plane (Illustration © by the author)

Long-Period Comets

Most newly discovered long-period comets are making their first appearance to human eyes, arriving in the inner Solar System after falling from the Oort Cloud on a journey that has lasted hundreds of thousands, or even millions, of years. After their brief showing most of them plunge back into the far depths of space, their orbits perhaps having been modified by gravitational interaction with the planets, but nevertheless remaining gravitationally bound to the Sun. It is possible that some comets arriving from the Oort Cloud will only undergo a single trip into the inner Solar System before being gravitationally perturbed into an ejection trajectory, where they escape the Solar System altogether and fly off into interstellar space. Only a few examples of such "single apparition" comets are known, C/2001 Q4 (NEAT) having been a nicely observed recent example. Even though there are undoubtedly 'rogue comets' in interstellar space—comets that have been ejected from their parent star systems. However, no observed comet has shown indisputable signs of belonging to such a class.

Short-Period Comets

With orbits of less than 200 years, it is now recognized that there are two categories of short-period comet—those of the Halley-type and those of the Jupiter family.

Halley-type comets (or, alternatively, 'intermediate-period' comets) have orbital periods between 20 and 200 years, and many move in orbits that are highly inclined (some inclined more than 90°) to the plane of the ecliptic. Thought to have originated in the Oort Cloud, their orbits have been altered by the gravitational influence of the planets (mainly Jupiter, Saturn, Uranus and Neptune) during their passage through the Solar System, shortening their aphelions (furthest point from the Sun in their orbit) to the extent that their escape from near-planetary space has been prevented. There are more than 60 currently known Halley-type comets, of which Halley's Comet, whose aphelion lies just beyond the orbit of Neptune, is of course the most well known.

Jupiter family comets have periods of less than 20 years, and most of them have inclinations of 30° or less. With a 3.3-year orbital period that takes it from inside the orbit of Mercury to inside that of Jupiter, Encke's Comet is the best known example, and like all Jupiter family comets its orbit is unstable and subject to considerable perturbation by the inner planets over time.

The origins of Jupiter family comets can be traced to the edge of the Solar System—the Kuiper Belt, the scattered disk and the zone of the centaurs.

Discovered in 1992, the Kuiper Belt dwarfs the main Asteroid Belt between Mars and Jupiter; extending beyond the orbit of Neptune to around 50 au from the Sun, it is 20 times as wide as the Asteroid Belt and is thought to contain up to 200 times its mass. While the Asteroid Belt contains objects made of silicates and metals, those of the Kuiper Belt come in the form of large chunks of frozen volatiles. There are perhaps as many as 100,000 Kuiper Belt objects (KBOs) larger than 100 km in diameter. More than a thousand KBOs are now known.

Gravitational scattering by the Solar System's four gas giants has produced a region beyond the Kuiper Belt populated by icy minor planets. Known as the scattered disk, it is far less densely populated than the Kuiper Belt, and its members have orbital eccentricities as high as 0.8, with inclinations up to 40° to the ecliptic. Within the scattered disk they never stray within 30 au of the Sun (the orbit of Neptune) but, being subject to gravitational perturbation by the outermost planet their orbits are unstable. As a result they are liable to either migrate into the zone of Centaurs inside the orbit of Neptune, shunted into the inner Solar System to become active comets, or flung out of the scattered disk to join the Oort Cloud.

Our final source of Jupiter family comets is the Centaurs, a class of SSSB whose orbits cross one or more of the outer planets. With paths that take them from perihelia beyond Jupiter's orbit out to the orbit of Neptune, their orbits are subject to considerable gravitational perturbation. There are 183 currently known Centaurs, one example being 2060 Chiron, the first member of the group to have been discovered, in 1977.

Chiron's classification as an asteroid was revised just a decade after its discovery, when it was found to have increased in brightness by 75 %, with follow-up observations revealing that it had developed a coma. Chiron is now classified as both asteroid and comet—and a hefty one at that, measuring 233 km in diameter—whose highly eccentric orbit takes it from just inside the orbit of Saturn to just outside Uranus' perihelion. Its orbit is liable to be perturbed so much that in around a mil-

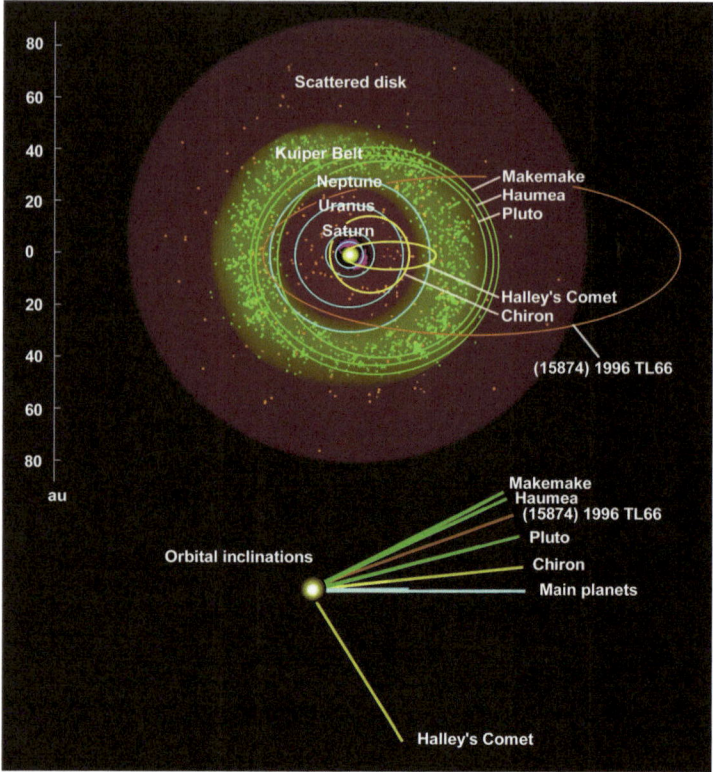

Fig. 1.11 The region of the Scattered Disk, Kuiper Belt and giant planets and the orbits (and orbital inclinations) of a select few of their constituent objects. Chiron was the first Centaur detected—now classified as a comet—while (15874) 1996 TL66 was the first Scattered Disk object found. Pluto, along with Makemake and Haumea, are giant icy worlds in the Kuiper Belt (Illustration © by the author)

lion years' time it will become a short-period comet. Cometary characteristics have also been found in other centaurs, including 60558 Echelus; like Chiron, on discovery (2000) it was initially thought to be an asteroid but later developed a coma (2005) (Fig. 1.11).

Main Belt Comets

Intriguingly, it has recently been discovered that the Asteroid Belt itself harbors a small number of objects that occasionally display comet-like activity (mainly during their perihelia). Main belt asteroids orbit largely between Mars and Jupiter, having an average distance from the Sun between 2 and 3.2 au and perihelia greater than 1.6 au. Main-belt comets are found to have near-circular, low

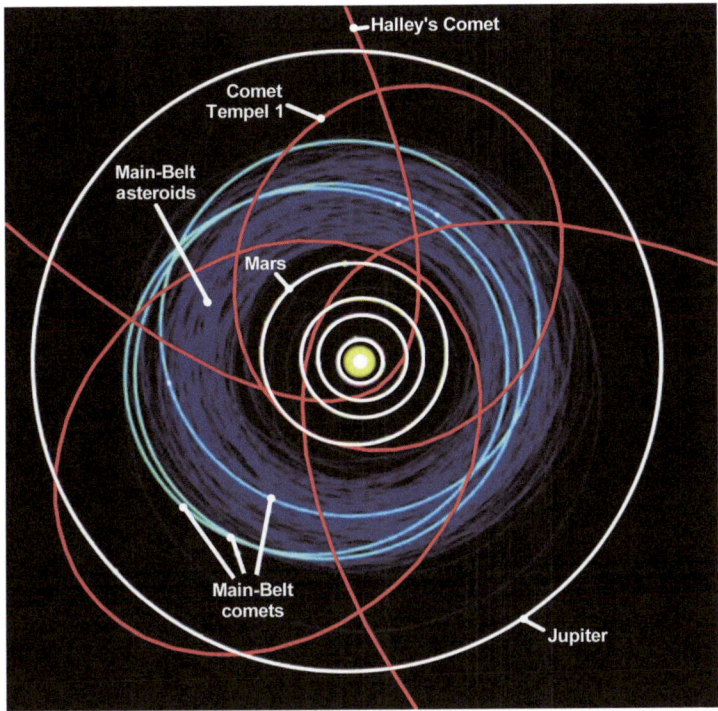

Fig. 1.12 Orbits of several comets lie entirely within the main Asteroid Belt (Illustration © by the author)

eccentricity and low inclination orbits, and most of them are orbiting in the outer part of the Asteroid Belt (Fig. 1.12).

Ten main-belt comets are currently known, although their nature is not currently well understood. They are likely to be in-situ primordial leftovers—icy asteroids, rather than true comets, that were somehow gravitationally levered into their current orbitally stable positions. As such, main-belt comets are likely to have silicate/metal cores overlain by icy deposits, or are an amalgamation of large silicate/metal fragments and ices. It is possible that many outer main-belt asteroids have such a makeup. Whatever their internal structure, their icy volatiles are well-hidden and it is only through sudden exposure by impact that the ices are exposed to sunlight, producing sublimation and the development of comet-like tails.

Asteroids: Vermin of the Skies

Until the latter half of the twentieth century astronomers recognized a clear distinction between comets and asteroids. Comets were considered to be relatively fragile icy objects orbiting the Sun in long ellipses in planes at all angles to the ecliptic, while asteroids were more solid rocky chunks largely confined to the main Asteroid Belt between Mars and Jupiter. That distinction has now changed dramatically; comets and asteroids are now considered to be part of a broad continuum of Solar System objects, lying midway between the smallest micrometeoroidal dust particles and the planets themselves.

Only 250 years ago, the existence of asteroids—independent planets so small that they would appear as specks of light through even the largest of telescopes—was completely unknown and largely unsuspected. As far back as 1596 Kepler had posited his belief that an as-yet unseen planet existed between Mars and Jupiter, a belief based on his conviction that a supreme being would not have wasted such a gap when creating the Solar System.

Taking a more scientific viewpoint, it is true that astronomers suspected that more undiscovered satellites awaited discovery around the planets themselves; between 1671 and 1684 *Cassini* had found Iapetus, Rhea, Tethys and Dione. But nobody had any good reason to speculate that there might be such tiny worlds in their own orbits around the Sun. No asteroid ever becomes bright enough to be easily seen with the unaided eye, and only one, Vesta, occasionally scrapes into naked-eye grade. It's hardly surprising, therefore, that nearly two centuries had elapsed after the invention of the telescope before the first asteroid was discovered.

In his *Contemplation de la Nature* (1764), Charles Bonnet wrote: "We know seventeen planets that enter into the composition of our Solar System (that is, major planets and their satellites); but we are not sure that there are no more." Oddly, he neglected to mention comets, although as we have seen he considered them to be planet-like members of the Solar System. Bonnet, along with most other astronomers of his era, knew very well that there were countless more comets at large in the depths of space; astronomers had become used to the fact that comets sped through the Solar System after suddenly appearing from what seemed to be a vast unseen reservoir far from the Sun.

In 1766 Bonnet's work was translated by the German astronomer Johann Titius (1729–1796), who added a simple mathematical formula that enabled the known planets' distances from the Sun to be derived. What made this formula intriguing was that it also appeared to predict the distances of other, as-yet undiscovered planets, suggesting that not only were there planets further out from the Sun than Saturn, but there was a place between Mars and Jupiter where a planet *ought* to be located. Titius' formula reappeared in a book by Johann Bode (1747–1826) and as a result is sometimes known as the Titius-Bode law, although this isn't a real law in the scientific sense.

According to the Titius-Bode law, 0 indicates the innermost planet Mercury, 3 for Venus, and each subsequent planet is then double the preceding figure from the Sun: 0—Mercury; 3—Venus; 6—Earth; 12—Mars; 24—unknown planet; 48—Jupiter; 96—Saturn; 192—unknown planet; 384—unknown planet; 768—unknown planet. Furthermore, if 4 is added to any of these planets and divided by 10, the result is that planet's distance from the Sun in Astronomical Units (au). One au is the average distance between the Sun and Earth, approximately 150 million kilometers.

Planet	T-B Law	T-B Law distance (au)	Actual Distance (au)
Mercury	0	0.40	0.39
Venus	3	0.70	0.72
Earth	6	1.00	1.00
Mars	12	1.60	1.52
Unknown planet	24	2.80	?
Jupiter	48	5.20	5.20
Saturn	96	10.0	9.54
Unknown planet	192	19.6	?
Unknown planet	384	38.8	?
Unknown planet	768	77.2	?

For several years following its publication, the Titius-Bode formula was regarded as a mathematical curiosity. Things changed in March 1781 when William Herschel (1738–1822) discovered Uranus; once its orbit was determined, the new planet was found to fit neatly into the Titius-Bode scheme—orbiting 19.2 au from the Sun, compared with 19.6 au as predicted by the formula—and it became a 'law' overnight.

Might there be yet more planets in the Solar System than had hitherto been suspected, worlds that had avoided detection because of their small size or low brightness? If the Titius-Bode law really did have predictive properties, then there ought to be a planet orbiting the Sun between Mars and Jupiter. Convinced that there was such a planet waiting to be discovered, yet was too small and dim to be easily seen, Franz Xaver (Baron von Zach, 1754–1832) organized the *Himmels Polizei* ('Celestial Police'), a group of 24 astronomers around Europe whose aim it was to systematically hunt down the mystery planet with a concerted telescopic search. Each 'officer' was to be allocated a section of the zodiac, a region on either side of the ecliptic, which would then be scrutinized for the smallest telescopic stars, measurements of stellar positions being compared with atlases.

On January 1, 1801—the very first day of the nineteenth century—Giuseppe Piazzi (1746–1826) made a fortuitous discovery from his observatory atop the Royal Palace in Palermo, Sicily. Piazzi, nominally a member of the Celestial Police (though it seems that he was never officially invited into its ranks), had been making corrections to a star catalog using his own observations. His technique was to observe each star for four nights before establishing its position, and to his surprise he discovered one in Taurus that had moved each night. Initially thinking that the new object might be an incoming comet, Piazzi made 24 measurements of its position over the following few weeks. News of the discovery quickly spread in the press and reached Jérôme Lalande (1732–1807) in Paris and Bode in Berlin by the end of February.

But Piazzi was reluctant to describe it as a new planet. He called it '*Ceres Ferdinandea*' in honor of the Roman goddess of agriculture and King Ferdinand

Fig. 1.13 Ceres and the Moon to scale. The image of Ceres was taken by the Hubble Space Telescope (Photo courtesy of STScI/Illustration by the author)

III of Sicily. In a letter to his friend, Celestial Police officer Barnaba Oriani (1752–1832), he wrote (Fig. 1.13):

> I have announced this star as a comet, but since it is not accompanied by any nebulosity and, further, since its movement is so slow and rather uniform, it has occurred to me several times that it might be something better than a comet. But I have been careful not to advance this supposition to the public. As soon as I shall have a larger number of observations, I will try to compute its elements.

Piazzi's measurements showed that the new object's path was less like that of a comet and more like that of a planet, an ellipse of low eccentricity aligned with the ecliptic. After losing sight of it in the glare of evening twilight, Piazzi set about calculating where it would be when it re-emerged in dark morning skies later that year; yet his attempts, and the efforts of others, to rediscover it failed. By July von Zach was getting pretty skeptical, and in a letter to Oriani he wrote:

> There are some astronomers who are starting to doubt the real existence of such a star. Burkhardt suspects that the observations are very wrong, it is a fact that he gave you and Bode a Declination wrong by at least half a degree, Burckhardt says that there are other

errors. Now I cannot conceive, how an observer as experienced as Piazzi, provided with the best instruments, a complete Circle, and a transit telescope by Ramsden, could incur such mistakes in his meridian observations?

Thinking that his Celestial Police force was becoming a laughing stock, von Zach's tone had become hostile, even sarcastic, by mid-December:

What is going on with the Ceres Ferdinandea? Nothing has been found as yet either in France or in Germany. People are starting to doubt: Already sceptics are making jokes about it. What is Devil Piazzi doing? Lalande wrote me that he [Piazzi] has changed again his observations and that he has made a new Edition of them! What does that mean? Lalande in his letter adds: This is why I do not believe in the planet.

In the end, *Ceres Ferdinandea* was recovered by von Zach himself on the last day of 1801, using fresh predictions made by a young mathematical genius, Carl Gauss (1777–1855); it turned out that von Zach had actually observed the object (but hadn't recognized the fact) before he had penned his recent tirade against Piazzi. However, the name *Ceres Ferdinandea* proved one word beyond the political pale for everyone's tastes, and it became known as Ceres. Amazingly, Ceres was found to orbit the Sun between Mars and Jupiter at an average 2.8 AU from the Sun, perfectly fitting into the Titius-Bode law. To honor its presumed planetary status it was to merit its own astronomical symbol, a shape resembling a sickle, similar to the symbol for Venus but with a gap in the circle.

Flushed with success, Gauss proceeded to work on a theory that took into account the motions of planetoids whose orbits were disturbed by the gravitation of large planets. His *Theory of motion of the celestial bodies moving in conic sections around the Sun* (1809) greatly simplified the mathematical tasks involved in the orbital prediction of asteroids and comets, since it required only three or more accurate positional measurements (right ascension and declination) of an object to be made. It remains a vital tool in astronomical computation and is even used in the software of GPS receivers.

Invigorated by the discovery of Ceres, and reasoning that there might be yet more minor planets to be discovered, the Celestial Police attempted to hunt them down with a passion matching that of their fellow comet-hunters. Further success came in March 1802 with the discovery of 2 Pallas by Heinrich Olbers (1758–1840) while observing Ceres; he went on to discover 4 Vesta in March 1807 (whose naming he granted to Gauss) and the periodic comet 13P/Olbers in March 1815. 3 Juno was discovered by Karl Harding (1765–1834) in September 1804.

William Herschel, famous for his discovery of Uranus, considered these new objects to be an entirely new class of entity. In an attempt to determine their size, he constructed a projection system and concluded that Ceres was 260 km across (underestimating its true diameter by almost four times) while Pallas was smaller, some 237 km across (less than half its known size). Clearly they were about as small as cometary nuclei, too small to be resolved into distinct disks, but showed no signs of a coma or tail. Orbiting in planet-like ellipses between Mars and Jupiter, in planes not too far off the ecliptic, their apparent brightness changed in such a way as to prove that they were solid objects reflecting sunlight; like the planets they were at their brightest at around opposition, when they appeared fully illuminated and closest to Earth.

Herschel pondered the question of what to call these objects, but in an attempt to protect the uniqueness of his own discovery, was unwilling to endorse any name that might give them a planetary status, such as 'planetule' or 'planeret.' Herschel even went so far as to finally settle on the word 'asteroid' (from the Greek, meaning 'star-like'), a word that neatly describes the telescopic appearance of these objects.

A promising start had been made in the new pursuit of asteroid discovery, but following the discovery of 4 Vesta in 1807 there was to be a gap of 38 years before any more were found. German amateur astronomer Karl Hencke (1793–1866) single-handedly revitalized the 'sport' in December 1845 with his discovery of 5 Astrea, followed in July 1847 by 6 Hebe (another name suggested by Gauss).

This renewed interest in asteroid discovery took place amid exciting times for astronomers. Planet Uranus' growing reluctance to obey the gravitational laws set down 150 years earlier by Isaac Newton posed one of the greatest problems facing astronomers at the time. English mathematician John Couch Adams (1819–1892) and French mathematician and astronomer Urbain Le Verrier (1811–1877) had been working on the very same problem, both convinced that Uranus' path was being influenced by the gravitational pull of a large unknown planet further out from the Sun. But where did their calculations suggest that this planet was, and once its position was known, who would be first to actually see it through the telescope?

In February 1844, George Airy (1801–1892), Astronomer Royal, learned of Adams' work on the problem and implemented a search for the planet. But Le Verrier had stolen a march on the English astronomers. On September 23, 1844, Johann Galle (1812–1910) at the Berlin Observatory, following Le Verrier's instructions, discovered Neptune through the observatory's 9-in. refractor. Incredibly, the planet was within just 1° of its predicted position. It shone too dimly to be seen with the unaided eye, and even through the telescope appeared as an incredibly small, pale blue disk.

With the Solar System's ever-increasing population, astronomy was becoming increasingly interesting. Spectacular C/1843 D1, known as the Great March Comet of 1843, proved to be frosting on the cosmic cake. One of the greatest comets in history, it was followed from Greenwich by John Hind (1823–1895), who later took charge of the South-Villa Observatory in Regent's Park. From here he discovered no fewer than ten asteroids, beginning with 7 Iris in August 1847 and ending with 30 Urania in July 1854. A prolific observer, Hind also discovered two comets during this period: C/1846 O1 (de Vico-Hind) and C/1847 C1 (Hind). Included among notable works of his that concern us is *The Comets: a Descriptive Treatise* (1852), his work in tracing the apparitions of Halley's Comet back to 11 BC, his careful analysis of ancient Chinese cometary records and his computations of the orbits of dozens of comets and asteroids.

By the mid-nineteenth century asteroid discoveries were coming in so thick and fast that giving each one its own special planetary symbol was becoming extremely impractical. Astronomer Johann Encke (1791–1865), famous for his computational work on the orbits of asteroids and comets, suggested that asteroids be indicated

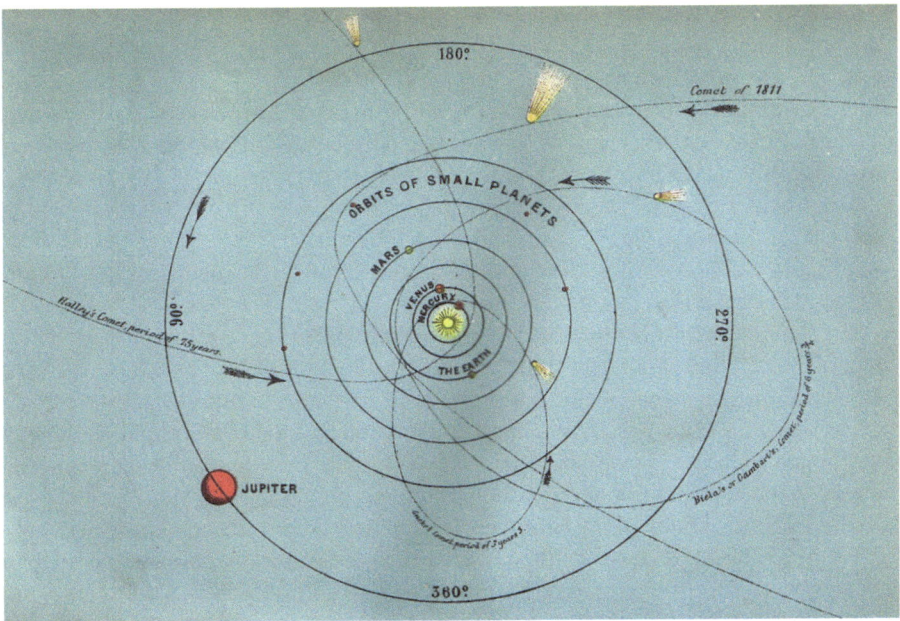

Fig. 1.14 An 1875 overview of the Solar System, showing the orbits of the planets, plus several well-known comets (Halley's Comet, Encke's Comet, Biela's Comet and the Great Comet of 1811) and the larger asteroids. From *Astronomy* by J. Rambosson (1875)

using an encircled number, and by 1860 the convention had become widely accepted. Around the same time the term 'minor planet' began to be used more frequently than 'asteroid' in the astronomical literature of Germany and France; the Royal Greenwich Observatory made the nomenclatural switch in the early twentieth century, while no firm consensus was ever reached among American astronomers. However, it must be noted that in 1947 the Minor Planet Center (MPC) was established in the United States under the auspices of the International Astronomical Union; in addition to asteroids, the MPC is responsible for the designation of all minor bodies in the Solar System, including comets (Fig. 1.14).

By 1886 no fewer than a hundred asteroids were known, 100 Hekate having been discovered by Canadian-American astronomer James Watson (1838–1880). Watson, whose tally of visual asteroid discoveries between 1863 and 1877 numbered 22, was among several prolific visual asteroid and comet discoverers of the era. The substitution of the keen eye by the photographic plate at the focus of large telescopes in the late nineteenth century saw a large rise in the numbers of asteroid discoveries. A time exposure made using an equatorially driven telescope shows stars as points, while any asteroids in the field betray their presence by producing tiny linear streaks as a consequence of their motion against the celestial sphere during the exposure. During the twentieth century, so many asteroids were being discovered photographically that they were sometimes rather unkindly referred to as 'the vermin of the skies.'

Meteors might be considered by some to be another form of celestial 'vermin.' Their bright streaks are capable of ruining a perfectly good photographic exposure of a comet or deep-sky object. Until the late nineteenth century it was thought that the meteoroids that gave rise to meteors when they burned up in Earth's atmosphere orbited the Sun in a single ring nearly coincidental with our planet's orbit. Their origin was unknown until Giovanni Schiaparelli (1835–1910) identified the relationship between meteor showers and comets.

In one of those cruel twists of fate that tend to distort the picture of the lives and works of scientists, Schiaparelli is largely remembered for his observations of nonexistent Martian 'canals.' But his most important contribution to astronomy was discovering the comet-meteor connection. In 1866 he found that orbital elements of the annual Perseid shower in August closely matched the orbital elements of a comet that had been discovered in July 1862—a comet now known as Comet 109P/Swift-Tuttle in honor of its discoverers, the American astronomers Lewis Swift (1820–1913) and Horace Tuttle (1837–1923). Schiaparelli pointed out that the Perseid stream is likely to be composed of trains of dusty debris flowing along the comet's elliptical orbit. He mathematically demonstrated that the comet and the members of the Perseid meteor shower crossed Earth's orbit in the same place, have similar directions of travel and follow the same course through space.

Biela's Comet and the Andromedid Meteor Shower

On 28 February 28, 1826, a Czech-born Austrian soldier and part-time amateur astronomer, Wilhelm von Biela (1782–1856), discovered a comet that was to astonish the scientific world. As the weeks went by, and news of the discovery spread, Biela's Comet (now officially designated 3D/Biela) became as bright as the third magnitude and was easily visible to the naked eye. The new comet was found to match one discovered by Charles Messier amid the stars of Eridanus in 1772, and another seen in 1805; it soon dawned upon astronomers that all three comets were undoubtedly the same object with a period of 6.6 years,

Biela's Comet came and went through our planetary neighborhood again in 1832–33, actually having been recovered in September 1832 by the famous English astronomer John Herschel. The comet's apparition of 1839 went unobserved because of the poor position of the Earth in relation to the comet, but it made a good showing at its next apparition, in 1845. Observers noted a peculiar 'antitail' streaming from the comet's head towards the general direction of the Sun.

In early January 1846 Matthew Maury (1806–1873) at the U. S. Naval Observatory was astonished to discover that the comet had split into two distinct pieces connected by a faint, hazy trail of dust and gas. The two components—each with their own fuzzy coma and small tail—were at first unequal in brilliance, but

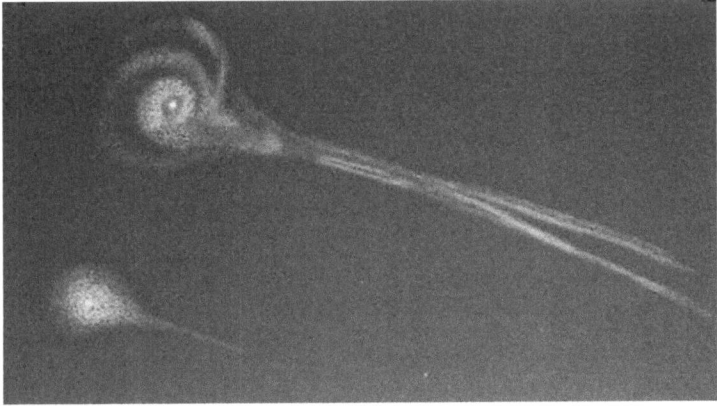

Fig. 1.15 Biela's Comet as it appeared in February 1846 after it had split into two pieces. From Bilderatlas der Sternenwelt (1888) by E. Weiß

over the next month the smaller companion brightened to become the other's equal. It then faded, and in March it completely disappeared, while its companion continued to be visible for another month or so. Joseph Hubbard (1823–1863) calculated the two objects to be around 340,000 km apart (Fig. 1.15).

On comet Biela's return in 1852, the pair had moved apart from each other to a distance of 2.6 million km. The two companions took turns at being the brighter, and it was not possible to determine which of them was the main body. The pair plunged into the depths of interplanetary space in September 1852 and have not been seen since. As the position of Earth and the comet was unfavorable in 1859, nobody was surprised that the object was not recovered, despite an intensive search. However, in the 1865–66 season Biela's Comet should have been easy to observe, but no trace of it was glimpsed.

Through the pioneering work of people such as Schiaparelli, astronomers were by now aware of the comet-meteor connection, and many predicted that in November of 1872 a shower would occur when Earth was in the vicinity of Biela's orbital path. In 1867 Edmund Weiss (1837–1917) of Vienna Observatory stated that, according to his calculations, a grand Andromedid meteor shower would take place on November 28, 1872. Weiss' prediction was a day out. On November 27 a magnificent meteoric display took place, with its radiant in the constellation of Andromeda, on the very same day that Earth intersected the path of Biela's Comet. If it was still out there, it would have been at that point in its orbit 2 months previously.

Although the shower was by no means as spectacular as the Leonid storm of 1866, many who saw it were of the opinion that it was better than any other they had ever seen, Leonids excepted. Edward Lowe (1825–1900) at Highfield Observatory near Nottingham, England, estimated that he had observed no fewer than 58,660 meteors between 17 h 50 min and 22 h 50 min. He remarked: "The striking feature (more especially in the earlier portion of the display) was the extreme smallness of the great portion of the meteors, not one in ten being equal to a star of the third magnitude, and many were as minute as the smallest visible stars and might aptly be called meteor dust."

The next Andromedid meteor spectacular was seen 13 years later, on November 27, 1885, a display that ranked as being equal to, if not better than, the shower of 1872. The whole event was observed under clear French skies by C. M. Vaison who, writing in *The English Mechanic and World of Science* (December 4), reported:

> Last night, 27th inst., we had here a most beautiful display of meteors, radiating from the neighbourhood of Gamma Andromedae. Hundreds of people went out of their houses to contemplate with very diverse feelings this beautiful phenomenon. Fear however was uppermost as some people, men and women, came to me to ask in voices altered by emotion, what all this meant; whether it was a sign of inundation, of war or pestilence, or whether we were all going to be squashed or burnt by the stars falling upon us. They were soon reassured, I am glad to say, by a few words of mine and many were their expressions of gratitude as they went away.

> Not having seen at the time the circular of Lord Crawford I did not expect this display myself. Soon after sunset I was observing Beta Delphini when I was surprised to see shooting stars crossing the field of my telescope. Other shooting stars, seen sideways, made me look up, and there I saw many large ones crossing various parts of the sky, their direction being at that time from east to west. No star was yet visible to the naked eye. As darkness came on, the whole expanse was crossed by these flying brands. Some were just blazing away as if gunpowder had exploded at that particular spot of the sky; others crossed the zenith for a short distance and many others bright and faint were falling from various heights in the heavens all around down to the horizon, but all seemed to have the same focus. Many of the larger meteors left long trails after them, lasting ten to twelve seconds, and then melting slowly away…At times the meteors were so numerous that it seemed as if all the stars were moving.

> The sky was perfectly clear from zenith to horizon all around; a strong dew was forming and we had a splendid glow at sunset. There is no doubt that this display began before sunset but by ten o'clock it was evident that we had nearly crossed the path of these meteors.

On the whole, British skies were more cloudy. Robert Grant (1814–1892) at Glasgow Observatory, Scotland, was lucky to have had clear skies that night. He'd also seen the previous Andromedid shower of 1872 and the Leonid star storm of 1866, and being in a unique position to compare all three events, Grant wrote:

> It has been my good fortune to have seen from this observatory the great meteor shower of November 13 1866, and also that of November 27 1872, in both instances under exceptionally favourable circumstances, and at the time of the recent shower I was naturally led to institute a comparison between the brilliant apparition then visible and the two preceding displays. The shower of the Leonids appeared to me to be beyond comparison the grandest of the three apparitions, both in respect to the abundance and the magnitude of the meteors. On the other hand, the recent apparition of the Andromedids offered a most striking resemblance in every respect to its brilliant predecessor of 1872.

It was during the great Andromedid shower of 1885 that the first photograph of a meteor was secured on a plate taken by Ladislaus Weinek (1848–1913), director of Praha Observatory in the Czech Republic. As an interesting aside, 1885 was notable for the discovery of the first extragalactic supernova, an exploding star in the Andromeda Galaxy whose light had taken 2.5 million years to reach the eyes of Victorian astronomers and became bright enough to be just visible with the naked eye. At the time of the Andromedid shower the supernova had faded to magnitude +13 and was visible only in moderate-sized telescopes. It's quite a coincidence that the two events were centered on the one constellation.

Another remarkable coincidence was the fall of a meteorite, on the night of the great Andromedid shower, at Conception Ranch 13 km east of Mazapil in Mexico. At the time the meteorite, a lump of iron weighing 4 kg, was only the ninth iron meteorite seen to fall, and the only one known to have fallen during the peak of a meteor shower. However, the object's initial path through the sky was not observed, so its radiant point is unknown. It is unlikely (though by no means impossible) that the Mazapil meteorite originated in the Andromedid stream. Eulogio Mijares, the rancher who recovered the object from a crater 30 cm deep, described the fall:

> It was about nine o'clock in the evening when I went to the corral to feed certain horses. Suddenly I heard a loud sizzling noise exactly as though something red hot was being plunged into water, and almost instantly there followed a somewhat loud thud. At once the corral was covered with a phosphorescent light and suspended in the air were small luminous sparks as though from a rocket. I had not recovered from my surprise when I saw this luminous air disappear, and there remained on the ground only such a light as made when a match is rubbed.

> A number of people from the neighbouring houses came running towards me, and they assisted me to quieten the horses which had become very much excited. We all asked each other what could be the matter, and we were afraid to walk in the corral for fear of getting burned. When in a few moments we had recovered from our surprise we saw the phosphorescent light disappear, little by little, and when we had brought lights to look for the cause we found a hole in the ground and in it a ball of light. We retired to a distance fearing it would explode and harm us. Looking up to the sky we saw from time to time exhalations of stars which soon went out but without noise.

> We returned after a little while and found in the hole a hot stone which we could barely handle, which on the next day looked like a piece of iron. All night it rained stars but we saw none fall to the ground as they seemed to be extinguished while still very high up.

Immediately following the 1885 Andromedid shower, astronomer Wilhelm Klinkerfues (1827–1884, the discoverer of six non-periodic comets) of Göttingen, Germany, realized that the comet responsible for the dramatic meteoric outburst might be nearby. He excitedly sent a telegram to his acquaintance Norman Pogson (1829–1891) in Madras, India. The telegram—transmitted via numerous relay stations and taking more than an hour and a half to reach its destination—stated: "Biela touched Earth last night. Search near Theta Centauri."

Pogson was able to make a search only after the weather had improved 2 days later. On December 2 he discovered an apparent cometary object that lagged behind the calculated position of Biela's Comet by 2 months. Pogson reported seeing the comet again the following evening, but as his observations were not corroborated by reliable sources, and the positions he gave were wide of the mark, the new comet could not be reconciled with either component of Biela's Comet or in any way considered to be a portion of Biela's nucleus that had mysteriously been delayed in its orbit.

Celestial mechanics often confound common sense. If an object in orbit around the Sun is slowed, then it will move towards the Sun, spiraling closer to it and actually speeding up its orbital period. A comet could apply its own unique 'retro-rockets' by means of renewed activity from its volatile icy nucleus, but any such 'eruption' of Biela's Comet would have had to have been pretty substantial to have given rise to such a substantial delay in arrival.

Records of the Andromedid meteor shower long precede the splitting up of Beila's Comet, having been traced back to AD 541. A great Andromedid shower was seen from Russia in 1741, and careful observations of the 1798 shower enabled two German astronomers, Heinrich Brandes (1777–1834) and Johann Benzenberg (1777–1846), to make the first calculation of a meteor's height above Earth. To do this, Brandes and Benzenberg set up observing stations several kilometers from each other, and simultaneously noted the positions of the tracks left by 22 meteors between September 11 and November 4. Because the distance between the observers was known accurately, and the observed locations of the meteors could be measured against the starry background, all that was needed was simple mathematics (a technique called parallax) to determine the real height of the meteors above the ground. Noting where the meteors appeared to become extinguished, most meteors were shown to have faded out at heights of between 10 and 214 km, giving a mean disappearance height of 89 km. Meteors were proven to be an atmospheric phenomenon likely to result from the infall of small extraterrestrial particles.

Perturbations by Jupiter have since diverted the Andromedid stream away from its intersection with Earth's orbit, and the once-busy Andromedid shower has now faded away.

Major Annual Meteor Showers and Their Cometary Progenitors

Quadrantids

In 1979, after analyzing observations of a comet recorded in 1490 by Chinese, Japanese and Korean astronomers, Iseo Hasegawa identified C/1490 Y1 as the progenitor of the Quadrantid stream. Peter Jenniskens of the Minor Planet Center has shown that minor planet 2003 EH is likely to be the extinct nucleus of C/1490 Y1.

Active between January 1–5, the year's first annual meteor shower reaches its maximum on January 3–4. For observers in the northern hemisphere, Quadrantid observing is practical from about midnight onwards, its circumpolar radiant in northern Boötes reaching a good elevation during the morning hours for observers in the northern hemisphere. The shower is not well seen from the southern hemisphere.

Meteor showers usually take their name from the constellation in which their radiant is located, but Quadrans won't be found on any modern star chart. It takes its name from a now-defunct constellation—Quadrans Muralis, the Wall Quadrant—located between Boötes and the tail of Ursa Major. ZHRs (zenithal hourly rates, an ideal based on how many meteors will be seen if the radiant is overhead on a moonless night) vary from year to year, averaging 60 but with some highs exceeding 120. Maximum usually takes place over a short, sharp time period. Medium-speed, Quadrantids encounter Earth at a velocity of 42 km/s, often producing distinctly blue meteors with fine trains.

The Lyrids

The Lyrids were spawned by C/1861 G1 (Thatcher), discovered visually by Albert Thatcher from New York City (when the Big Apple enjoyed considerably darker night skies).

The shower makes a tentative start in mid-April, hitting their peak on April 22. Rates of around 10–15 meteors per hour can usually be expected under dark conditions at their peak. Lyrids are bright and fast, with a velocity of 48 km/s; most are train-free, and they produce a fair number of fireballs. From mid-northern latitudes observing can begin an hour or 2 before midnight, when the radiant (in eastern Hercules, southwest of Vega in Lyra) has risen well above the northeastern horizon. Lyrid rates on the nights either side of maximum average around 5 per hour, with lower rates during the rest of the shower's period of activity until April 25.

Eta Aquariids

1P/Halley (Halley's Comet) is the Eta Aquariids' progenitor—a distinction that it shares with the Orionids of October. Activity from the Eta Aquariids takes place from April 19 to May 26, peaking on May 5–6. It is one of the more difficult showers to observe from the northern hemisphere because of the low altitude of its radiant, plus the fact that the radiant is visible only for a few hours above the horizon before dawn. Observers in the southern hemisphere enjoy a much better view of the shower, with dramatically higher rates between 10 and 30 per hour at maximum. Traveling at 67 km/s, Eta Aquariids are fast, and some 30–50 % of them leave a persistent train (Fig. 1.16).

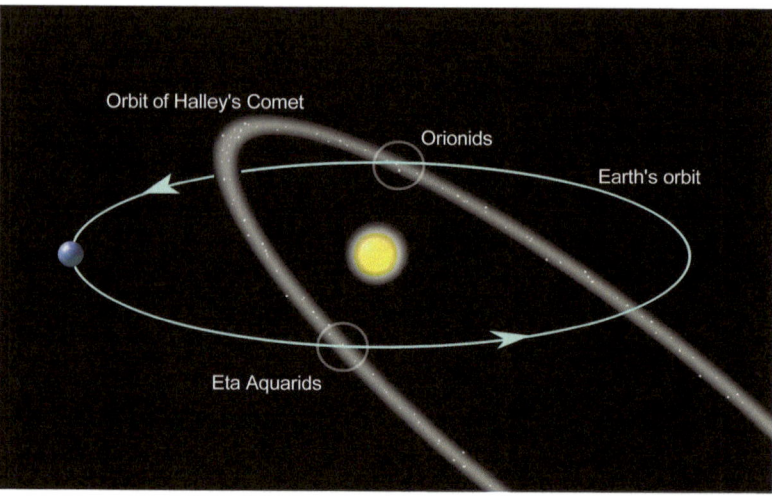

Fig. 1.16 Illustrating the mutual origin of the Eta Aquariids and Orionids, both belonging to the stream of meteoroids deposited by Halley's Comet (Illustration © by the author)

The Delta Aquariids

96P/Machholz, a comet discovered through binoculars by California amateur Don Machholz in 1986, is thought to be the Delta Aquariids' likely progenitor. After an exhaustive spectral comparison with 150 comets, a cometary composition program run by the Lowell Observatory found that 96P/Machholz has an unusual chemical composition, having significantly less carbon and cyanogen. This, in addition to being the only short-period comet with both a high orbital inclination and eccentricity, has led to speculation that the comet may be of extrasolar origin; having formed around another star system, it was gravitationally perturbed, flung out and subsequently captured by the Sun. With a period of 5.2 years, the comet came to perihelion in July 2012 when it was found to have split to produce two tiny fragments traveling ahead of the main nucleus. It next comes to perihelion on October 27, 2017.

The Delta Aquariids are another shower best viewed from the southern hemisphere. Active between July 21 to August 23, their peak of activity takes place on July 27–28, but good rates can be expected for around 3 days either side of this date. The shower produces faint, medium speed (42 km/s) trail-less meteors, and fireballs are rare.

The Alpha Capricornids

Discovered in March 2002 by NEAT (Near-Earth Asteroid Tracking Program) and designated minor planet 2002 EX12, images taken in July 2005 showed that the 'asteroid' had developed a tail. Hastily renamed 169/P NEAT, the comet has an eccentric orbit with an orbital period of 4.2 years. It was identified as the progenitor of the Alpha Capricornid meteor shower.

Active between mid-July and mid-August, the Alpha Capricornids have a maximum plateau centred on July 29. With a velocity of 24 km/s, they are slow moving and often yellowish in color. Maximum rates are very low, rarely exceeding five per hour; however, the shower produces an inordinate number of fireballs.

The Perseids

As we've already seen, in 1866 Schiaparelli proved beyond reasonable doubt that comet 109P/Swift-Tuttle is the parent comet of the Perseid meteors. Currently locked into a 1:11 orbital resonance with Jupiter, the comet has an orbital period of 133 years (11 times longer than one Jupiter orbit). With a nucleus measuring around 26 km across, 109P/Swift-Tuttle is the largest object in the Solar System known to make repeated close flybys of Earth. Although its orbit is known to be stable and having no impact danger over the next 2,000 years, astronomers' inability to be sure about the comet's behavior after a very close approach in 4479 has led to it being described as the single-most dangerous threat known to humanity.

Hitting Earth's atmosphere at a fast 60 km/s, usually bright with trains, the Perseids are active from mid-July to late August. Fairly reliable displays of at least 50–75 meteors per hour may be observed at the Perseids' maximum on August 12–13, indicating that the meteoroids are pretty evenly spread through space. This in turn suggests that the shower is very ancient; over a long period of time, the countless billions of particles with their slightly different speeds have dispersed around the entire elliptical orbit (which goes well beyond Pluto's orbit) of 109P/Swift-Tuttle. The first European records of the Perseids date back to 830, but the core of the steam may have been laid down around 25,000 years ago.

The Draconids

Although not the strongest of the year's meteor showers, the Draconids produced one of the twentieth century's most spectacular meteor displays. The outburst took place in European skies on the evening of October 9, 1933, over a period of 4 h. During the period of maximum activity, observers at Potsdam in Germany counted 5,700 meteors in 30 min. From Malta, Rupert Forbes-Bentley observed 22,500 meteors that night, reporting 480 in 1 min at the peak of the display. The brightness range of this splendid flourish of meteors went from that of Venus (magnitude −3) down to that of the limit of visibility. Another significant Draconid outburst was also observed in 1946, and material shed that year gave rise to another enhanced display in 2005.

In 1915 Martin Davidson (1880–1968) had identified the parent of the Draconids as 21P/Giacobini-Zinner, a comet with a period of 6.6 years that was discovered in 1900. From time to time the Draconids have been referred to in some astronomical publications as the 'Giacobinids.' The stuff making up the Draconid stream was directly sampled on September 11, 1985, when the joint NASA/ESA space probe International Cometary Explorer (ICE) flew into the coma of Comet Giacobini Zinner and came within 8,000 km of its icy nucleus. The probe (originally launched as ISEE in 1978 to observe the solar wind) had no dust shields like the Giotto Halley probe had, and as it plunged into the coma, meteoritic impacts hit the probe at a rate of one per second, proving the coma to be far less dusty than expected. ICE took 20 min to cross the comet's 24,000-km-wide tail and emerged more or less intact and in good working order. ICE's measurements of dust, molecules of water-ice and carbon dioxide, confirmed the theory that comets are large 'dirty snowballs.'

The Draconids are active between October 6 and 10, with a maximum on October 8. Their rates vary considerably, but are worth watching because of the possibility of another Draconid spectacular. With a velocity of just 20 km/s, they are very slow-moving meteors, and from northern temperate locations the radiant is circumpolar, highest before midnight.

The Orionids

Like May's Eta Aquariids, the Orionids' parent comet is 1P/Halley. However, neither stream is perfectly coincidental with the current orbit of Halley's Comet, nor are they coincident with each other. Earth encounters the Orionid stream some 0.155 AU from Halley's current orbit, while there's a difference of 0.066 AU for the Eta Aquariids. These are substantial differences, well beyond those displayed by most other meteor showers and their progenitors. It was only in the latter half of the twentieth century that astronomers had done enough research on the likely evolution of the orbit of Halley's Comet and the observed characteristics of the showers over time that a consensus was arrived at. From time to time Halley's Comet has entered into orbital resonances with Jupiter. It is thought that a 1:6 resonance with Jupiter (one orbit of Halley for every six Jovian orbits) took place between 1404 and 690 BC, followed by a 2:13 resonance between 240 BC and AD 1700. Outbursts of activity during these periods have produced dense concentrations of meteoroids, which in turn orbit in resonance with the giant planet. Although these meteoroid clumps follow the same orbit, the comet's orbit eventually changes and comes out of resonance. Whenever Earth encounters these denser meteoroid clumps, enhanced Orionid activity is observed. For example, the Orionid outburst in 1993 is thought to have been due to an encounter with 2:13 a resonant patch of meteoroids laid down in 240 BC; this may be repeated in 2070.

Owing to their high radiant the Orionids are much more favorably observed from the northern hemisphere than the Eta Aquariids. Active between mid- to late

October, they peak on October 21–22, usually giving rise to ZHRs of around 25. Orionids are renowned for their speed—with a velocity of 67 km/s they are the second fastest meteors after the Leonids.

The Taurids

The shower is composed of two close, related streams—the Southern Taurids and the Northern Taurids. Their common progenitor is comet 2P/Encke, whose outbursts 4,700 years ago and 1,500 years ago produced the streams.

Activity from the Southern Taurids takes place over more than 2 months, from early September to mid-November, with a very muted maximum on October 9–10, whose rates rarely exceed 5 per hour. Slow meteors with a velocity of 28 km/s, the Southern Taurids are noted to be rich in fireballs.

The Northern Taurids are active between mid-October and early December, peaking on November 12–13. Medium-speed at 30 km/s, they are often white and yellow, a small proportion of them appearing orange. From mid-October to mid-November, when both northern and southern activity is occurring simultaneously, there is often a rise in the number of observed fireballs. Owing to the closeness of their radiants in Taurus, it can be difficult to identify which of the showers a meteor belongs to.

The Leonids

For a few days each November Earth encounters a portion of the meteoroid stream deposited by comet 55P/Tempel-Tuttle, producing perhaps the most historically significant of all the annual meteor showers—the Leonids. From time to time this shower has displayed intense outbursts of activity known as meteor storms, where the frequency of meteors is inordinately higher than the shower's usual rates, amounting to hundreds or sometimes thousands of meteors per hour.

The first record of a Leonid storm dates back to 902, when a Moorish account from north Africa stated "[T]hat night there were seen, as it were lances, an infinite number of stars, which scattered themselves like rain to right and left, and that year was called the Year of the Stars." More recent Leonid storms have taken place in 1799, 1832, 1833, 1866 and 1966.

Meteor Shower of 1833

The first modern account of a Leonid meteor storm was given by Alexander von Humboldt (1769–1859), a famous German naturalist who happened to be on an expedition to South America during the storm, which took place on the morning of November 12. Von Humboldt observed the event from the Venezuelan village of

Cumana (300 km east of Caracas), writing: "At about 2.30 the most extraordinary, luminous meteors began rising out of the sky from the east and northeast...[A] lmost all the inhabitants of Cumana witnessed these phenomena because they had left their houses before four o'clock to attend early morning mass."

Humboldt remarked that within quarter of an hour after the storm began the sky was full of meteors, crowded so much that he reported "thousands of meteors and fireballs moving regularly from north to south with no parts of the sky so large as twice the Moon's diameter not filled each instant with meteors." Many were as bright as Jupiter and emitted showers of sparks as they flew across the sky, and some of them exploded violently in a spectacular terminal flare. Incredibly, Humboldt reported the storm was visible in the morning skies half an hour after local sunrise.

Andrew Ellicott (1754–1820), a U. S. government surveyor, observed the same storm from the deck of a ship off Cape Florida in the Gulf of Mexico. He recorded the event in his journal:

> About three o'clock am. I was called up to see the shooting of the stars (as it was com-
> monly called). The phenomenon was grand and awful; the whole heavens appeared as if
> illuminated with sky rockets, which disappeared only by the light of the Sun after day-
> break. The meteors, which at any one instant of time appeared as numerous as the stars,
> flew in all possible directions, except from the earth, toward which they all inclined more
> or less; and some of them descended perpendicularly over the vessel we were in, so that I
> was in constant expectation of their falling upon us.

During this event the concept of a radiant—a single small area of the sky from which meteors appear to emanate—became apparent. The 1833 Leonid storm sparked modern interest in the scientific study of meteors. American observers described the meteors as thick as snow coming down in a snowstorm. At Greenwich, England, eight observers were each allocated an area of sky, allowing complete coverage. A total of 8,000 meteors were recorded, with a maximum of 4,860 falling between 1 and 2 am.

Denison Olmsted (1791–1859) of Yale University claimed that the early morning of November 13, 1833, was:

> ...rendered memorable by an exhibition of the phenomenon called shooting stars, which
> was probably more extensive and magnificent than any similar one hitherto recorded ... The
> firmament was unclouded; the air was still and mild; the stars seemed to shine with more
> than their wonted brilliancy...Probably no celestial phenomenon has ever occurred in this
> country...which was viewed with so much admiration and delight by one class of specta-
> tors, or with so much astonishment and fear by another class. For some time after the occur-
> rence, the Meteoric Phenomenon was the principal topic of conversation in every circle.

Olmsted and others observed that the Leonid radiant remained fixed within the 'sickle' asterism of the constellation of Leo over a period of several hours during the storm, in spite of the fact that Earth had rotated to the east through at least 30°. This led Olmsted to the correct conclusion that the meteors must have been coming from an interplanetary, rather than an atmospheric, source—a conclusion that, surprisingly, was not endorsed by some of the leading astronomers of the day. Olmsted reported one observer had actually detected the Leonid stream before the meteor storm had commenced purely by the reflection of sunlight from the dense meteoric clouds at the stream's core.

Meteor Shower of 1866

55P/Tempel-Tuttle, the Leonids' parent comet, was independently discovered on December 19, 1865, from Marseilles, France, by Ernst Tempel (1821–1889) and on January 6, 1866, from the U. S. Naval Observatory, Washington, by Horace Tuttle. By February 9, 1866, the comet had faded from view. Calculations by Theodor von Oppolzer (1841–1886) showed that the comet's orbital period was 33.17 years and that it intersected with Earth's orbit. Later that year, a remarkable Leonid storm was observed. Sir Robert Ball, an eminent astronomer and popularizer of astronomy, gives an account of it in his classic *Story of the Heavens*:

> Such was the occurrence which astonished the world on the night between November 13th and 14th, 1866. We then plunged into the middle of the [Leonid] shoal. The night was fine; the Moon was absent. The meteors were distinguished not only by their enormous multitude, but by their intrinsic magnificence. I shall never forget that night. On the memorable evening I was engaged in my usual duty at the time of observing nebulae with Lord Rosse's great reflecting telescope. I was of course aware that a shower of meteors had been predicted, but nothing that I had heard prepared me for the splendid spectacle so soon to be unfolded. It was about ten o'clock at night when an exclamation from an attendant by my side made me look up from the telescope, just in time to see a fine meteor dash across the sky. It was presently followed by another, and then by others in twos and in threes, which showed that the prediction of a great shower was likely to be verified. At this time the Earl of Rosse joined me at the telescope.

The two astronomers made their way to the top of one of the high walls that surrounded the telescope and chose to view the event from there, and

> ...for the next two or three hours we witnessed a spectacle which can never fade from my memory. The shooting stars gradually increased in number until sometimes several were seen at once. Sometimes they swept over our heads, sometimes to the right, sometimes to the left, but they all diverged from the east...all the tracks of the meteors radiated from Leo. Sometimes a meteor appeared to come almost directly towards us, and then its path was so foreshortened that it had hardly any appreciable length, and looked like an ordinary fixed star swelling into brilliancy and then as rapidly vanishing. Occasionally luminous trains would linger on for many minutes after the meteor had flashed across, but the great majority of the trains in this shower were evanescent. It would be impossible to say how many thousands of meteors were seen, each one of which was bright enough to have elicited a note of admiration on any ordinary night.

Urbain Le Verrier used observations of the 1866 Leonids to compute a 33.25 year orbit for the Leonids. The closeness between the orbit of the Leonids and that of 55P/Tempel-Tuttle struck a number of astronomers, including Christian Peters, Giovanni Schiaparelli and von Oppolzer. The Leonid storms of 1799, 1833 and 1866 convinced astronomers that great displays occurred on a cycle matching close approaches by 55P/Tempel-Tuttle (Fig. 1.17).

Meteor Shower of 1899

It was with great expectations that astronomers faced the night of November 14, 1899. The interest of the general public, too, had been greatly aroused by newspapers,

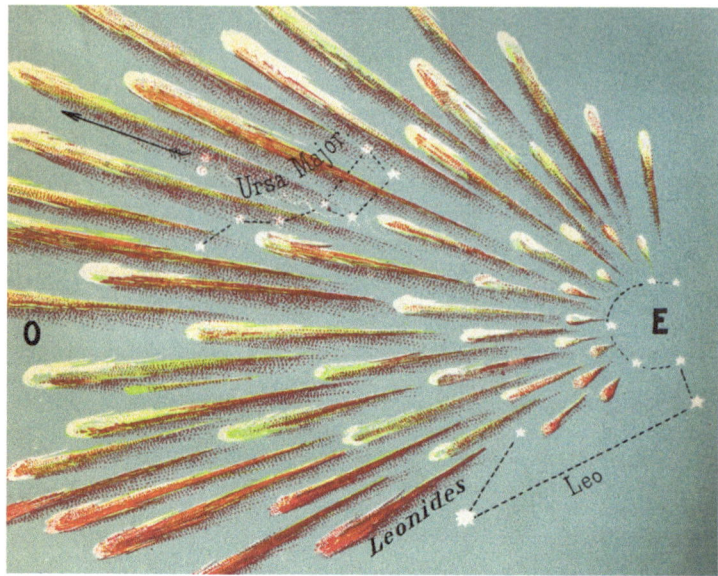

Fig. 1.17 Impression of a Leonid meteor storm. From *Astronomy* by J. Rambosson (1875)

keen to promote the forthcoming Leonid shower as a not to be missed display of celestial pyrotechnics. However, calculations made some time earlier by British astronomers George Stoney (1826–1911) and Arthur Downing (1850–1917) suggested that the densest part of the Leonid swarm would not be encountered by Earth on this occasion, as the meteoroid swarm had likely been gravitationally perturbed by Saturn in 1870 and Jupiter in 1898; unless the perturbed swarm was very wide indeed, then a major meteor display was unlikely to occur on this occasion. But few people took any notice of these negative predictions.

There was great disappointment when the anticipated Leonid storm of 1899 failed to happen. The 1899 Leonids did put on a display, though a very muted one, but they failed to impress most of the public who made the effort to observe them. Their disappointment was described by American meteor expert Charles Olivier (1884–1975) as "the worst blow ever suffered by astronomy in the eyes of the public, and has indirectly done immense harm to the spread of science among our citizens."

Meteor Shower of 1966

For almost a century after it had been lost to view in 1866, astronomers had assumed that comet 55P/Tempel-Tuttle had disintegrated, and following the Leonid disappointment of 1899 there was a general lack of interest in making a special effort to observe the shower. However, in 1965 calculations by German astronomer Joachim Schubart produced a new plot of the comet's orbit, enabling astronomers to rediscover it. Working back, it was found that 55P/Tempel-Tuttle had been observed in 1699 by Gottfried Kirch (but hadn't been recognized at that time as a periodic comet).

Calculations showed that in 1965 comet 55P/Tempel-Tuttle had passed just 0.0032 AU from Earth's orbit—its closest approach since 1833. There was an intense renewal of interest in the possibility of another Leonid storm, but nobody was sure where on Earth the burst of meteors would be seen from.

On November 16–17, 1966, the Leonids put on the twentieth century's greatest display of shooting stars, as meteors fell 'as thick as snowflakes.' The storm was visible in the morning skies between western parts of the United States and eastern Asia. The storm reached its peak at 12 h UT on November 17, when the rate was estimated to be about 150,000 naked-eye meteors per hour. It was calculated that the meteors passed over Arizona at a rate of 2,300 each minute in a period of 20 min from 5 am local time on November 17. Observers in 1966 commonly reported the illusion of traveling rapidly through a tunnel lined with stars.

Modern research exploring the relationship between 55P/Tempel-Tuttle and the Leonids has revealed a complex picture. In 1981 Donald Yeomans of the Jet Propulsion Laboratory in California examined observational data for the Leonids from 902 to 1969; mapping the likely dust distribution surrounding the comet. He found that most of the ejected dust lagged behind the comet and was outside its orbit. This indicated that the solar wind, in addition to gravitational perturbations by the planets, were responsible for changes in the evolution of the Leonid stream.

More recent research on the links between comets and their associated meteor showers—notably by V. V. Reznikov and E. A. Emel'yanenko (Kazan University), David Asher (Armagh Observatory), Robert McNaught (Siding Spring Observatory) and Esko Lyytinen (Finnish Fireball Working Group)—has produced great break-throughs. Individual streams from each passage of the comet have been identified, and detailed account has been taken of the specifics of their parent comets and their orbits, rates of nucleus rotation, particle size and ejection velocities, in addition to the role of solar radiation pressure in differentiating particle sizes in the ejected streams. This has enabled the intensity of each storm, due to variations in the distribution of meteoroids, to be narrowed down and made more predictable.

As a consequence it is now understood that the Leonid storm of 1833 was due to an encounter with a dense trail left in 1800, while the 1866 storm was produced when Earth passed through an even older clump of debris—one deposited during comet 55P/ Tempel-Tuttle's 1733 visit. Proving the old saying that good things come to those who wait, Earth encountered a dense trail left in the wake of the comet's 1899 passage during the 1966 Leonid storm. Two rich bursts of Leonid activity took place in 2001 and in 2002—both with double peaks, enhancements caused by dust ejected by 55P/Tempel-Tuttle in 1767 and 1866.

Enhancements of Leonid activity will continue to take place throughout the twenty-first century, sometimes producing good displays, especially in 2034, 2035, 2061, 2069 and 2094, the latter possibly including the best Leonid storm of the century.

What of the Leonid shower itself? The multitude of particles that burn up in Earth's atmosphere appear to radiate from an area of sky in the constellation of Leo within the sickle shape of stars that make up the lion's head and flowing mane. The Leonid meteors are visible every year between November 5 and 30, with most activity occurring on November 17–18. Moving at a velocity of 71 km/s, they are the swiftest meteors of all the annual showers. Many leave behind faintly glowing trains that can persist in the atmosphere for a few moments after the meteor's bright flash has disappeared. The normal rate for Leonids lies in the region of 5 to 20 meteors each hour during maximum.

The Geminids

Asteroid 3200 Phaethon is the parent body of the year's most consistently strong meteor shower. The first asteroid to be found on spacecraft imagery, Phaethon was discovered by Simon Green and John Davies on images secured by the Infrared Astronomical Satellite (IRAS) on October 11, 1983. Classed as an Apollo asteroid, its comet-like orbit brings it nearer the Sun than any other named asteroid, with a perihelion of 0.140 AU, making it a Mars-Earth-Venus-Mercury-crosser. After its orbital elements were determined, astronomers realized that they matched the Geminid meteors.

About 5 km in diameter, Phaethon is a B-type asteroid in a high inclination orbit, a primitive carbonaceous body rich in volatiles, part of the Pallas family of asteroids that are thought to be remnants of the early Solar System. There is strong reason to suspect that Phaethon, like other objects in the Pallas family, is an impact fragment from 544-km-diameter Pallas itself.

So unlike a typical comet in substance, Phaethon posed a problem for astronomers. How could it produce meteors, given that it moves so close to the Sun, making the continued existence of sub-surface ices unlikely? Studies showed that dense patches within the Geminid stream arose from ejection when Phaethon has been near perihelion. Then, on June 20, 2009, astronomers were surprised to observe Phaethon suddenly brighten by around ten times its normal magnitude. Having taken place when the asteroid was near perihelion, the brightening is thought to have been due to a release of dust after surface rocks cracked due to heating, dust being raised above the surface by electrostatic levitation (where electrons in the solar wind charges dust that then repels itself) and is then carried off by the radiation pressure of sunlight.

The idea of a 'rock comet' is in some respects appealing, but it does not explain how some of the heftier Geminid meteoroids—some the size of softballs, far too large to have been electrostatically lifted off the surface—are produced.

Although the structure and composition of Phaethon remains something of a mystery, the Geminid meteors produce a reliable annual shower that's active between December 7 and 16. At their peak on December 13–14, their rates can easily exceed 75 per hour under ideal conditions, and they are best observed from the northern hemisphere where the radiant is visible all night, highest after midnight. Slow-moving at 35 km/s, the Geminids produce plenty of bright meteors, many of which are yellowish in color.

Chapter 2

From Ancient Atmospherics to Icy Dirtballs

Identity Crisis

Comets have an identity crisis of historic proportions. Like so much in astronomy, appearances alone can be deceptive. A bright comet in the twilight sky—replete with dazzling head and glowing, sky-spanning tail, its position and appearance changing each night—looks as though it's making a concerted effort to be the most important object in the night skies. As we'll find out, that grand, jaw-dropping cometary spectacle is largely the result of celestial 'smoke and mirrors.' A great comet is one of nature's most astonishing special effects.

Complete ignorance may have been acceptable 500 years ago, before the age of scientific inquiry, when nobody knew anything about comets other than they appeared in the skies without warning, they were very far away and they could assume a variety of shapes and sizes. There was also a deep, primeval overlay of fear and superstition—common among varied cultures the world over—that painted comets in a pretty bad light (Fig. 2.1).

Incredibly, hundreds of years of scientific investigation have still not completely vanquished age-old comet phobias; indeed, the use (or should we say misuse) of scientific discoveries and the rise of pseudoscience has seen the perpetuation of comet fears and the appearance of new anxieties. Regrettably, there's nobody more deceived than the average 'person on the street,' whose information about space subjects may be derived from a mixture of cinematic fiction and media sensationalism.

In addition to the convincing illusion of cosmic grandeur and the mire of old and new superstition surrounding comets, another manifestation of the cometary identity crisis is the changing scientific picture of the physical nature of comets and

P. Grego, *Blazing a Ghostly Trail: ISON and Great Comets of the Past and Future*,
The Patrick Moore Practical Astronomy Series, DOI 10.1007/978-3-319-01775-4_2,
© Springer International Publishing Switzerland 2014

Fig. 2.1 The apparition of Halley's Comet in 1066, famously portrayed in the Bayeux Tapestry as an ill omen for King Harold prior to the Norman conquest of England

where they come from. Of course, changing pictures is part and parcel of the essence of science itself. The story of the astronomical investigation into comets, from the speculations of the ancient Greeks, through the work of Edmond Halley to close-up studies by space probes, is truly fascinating.

Cosmic Speculations in Ancient Greece: One Comet Fits All

Philosophers of ancient Greece inquired into the very nature of the cosmos. They were the first to make serious attempts to explain celestial objects and phenomena—from the very small to the very large, from the basic atomic essence of matter to the structure of the universe—without necessarily calling into action the hand of supernatural entities. Comets, too, were the subject of considerable speculation.

There was no overall consensus in ancient Greece on how the cosmos was thought to be arranged. Philosophers were free to speculate, and a number of schools were founded, each dedicated to certain models and ways of thinking. Early Greek mythology fancied that Earth was a circular disk comprising a central landmass surrounded by a great ocean; above was the air, then the cosmic ether, while below Earth lay Hades. Comets were perceived as signs of the gods and portents of things that were to transpire on Earth, indicators of terrestrial change and signposts to the fate of societies and individuals.

By the fifth century B.C., however, philosophers had largely abandoned this simplistic picture of the universe as civilization flourished and practical problems needed to be addressed by some serious thinking. For example, Greek seafarers navigating by the stars found that new stars appeared above the southern horizon as they voyaged further south. Clearly, Earth was not flat, and the vault of heaven wasn't just decorated with a fabulous tapestry pinned up by the gods just for mortals to admire.

It's possible here to cite only a select few ancient Greek speculations on the nature of the heavens and to give just the bones of these theories. Thales of Miletus (c. 624–546 B.C.)—recognized by many to have been the first great Greek philosopher—suggested that everything in the cosmos was made up of 'moisture,' a universal substance that can variously manifest itself as air, water, fire, earth and ice.

After traveling widely around the Middle East, a school was founded in Samos by the great mathematician and astronomer Pythagoras (c. 570–495 B.C.). He taught that it was possible to understand the universe through mathematics and held that the circle and the sphere were perfect figures. The number ten was particularly important to Pythagoreans; ten, the tetractys, was the very number of the universe. Early Pythagoreans held that Earth was at the center of the universe. Surrounding it was a nested series of transparent shells—revolving crystalline spheres—upon which celestial objects were attached. Stars were attached to the outermost sphere, beyond which lay an infinite void. Inside lay spheres that carried (in order of decreasing distance from Earth) Saturn, Jupiter, Mars, Venus, Mercury, the Sun, the Comet (a single object like a planet), the Moon and Earth.

An elaborate cosmological model was created by the Pythagorean philosopher Philolaus (c. 470–400 B.C.), who replaced a central Earth with a Central Fire, the powerhouse of the universe. Naturally, there was a total of ten spheres. Antichthon, the 'Counter-Earth'—a body that was permanently hidden on the other side of the Central Fire—occupied one of these spheres. Antichthon was a peculiar concept introduced to give 'balance' to the system, since Earth itself seemed to be such a heavy object compared with the planets. As for the very hub of the cosmos, the Central Fire, Earth revolved so that the hemisphere opposite Greece was always turned towards it. Setting all these objects in their various rates of motion very loosely approximated the real motions of the universe, but the actual motions didn't hold up against any sort of rigorous comparison with observations. For example, Pythagoreans considered that the Comet was one of the planets, but that it appeared at great intervals of time and only was a little above the horizon; because of this, some apparitions of the Comet failed to be seen owing to their low altitude (Figs. 2.2 and 2.3).

Hippocrates of Chios (c. 470–410 B.C.), not to be confused with his contemporary Hippocrates of Cos, known as the father of western medicine, wrote one of the first textbooks on geometry and, influenced by the Pythagorean school, speculated on the nature of the cosmos. As an aside, it's interesting to note that Hippocrates also came up with the striking notion that light rays came from the eyes to illuminate

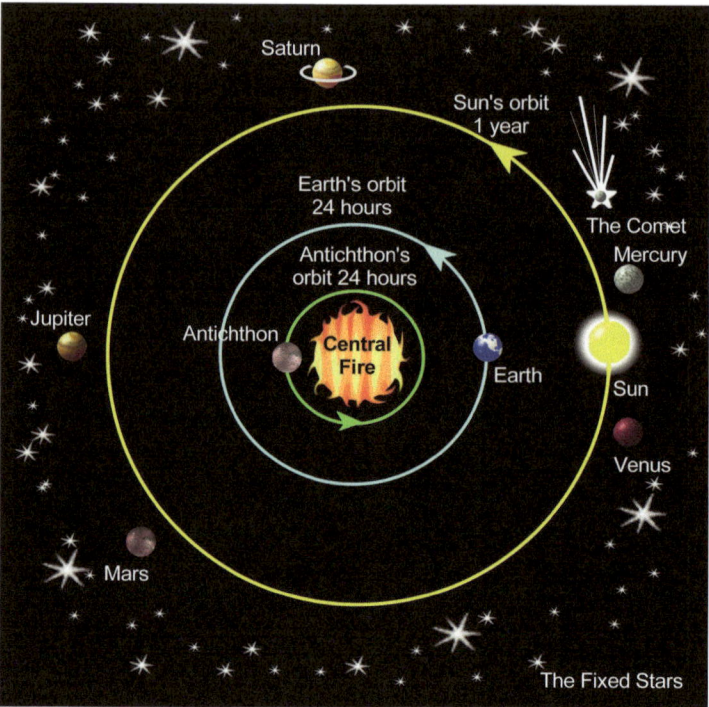

Fig. 2.2 Philolaus' model of the universe, with its Central Fire, Antichthon (Counter-Earth) and elusive Comet (Illustration by the author)

objects, rather than the other way around, which just goes to show how creative his thinking could be.

Like the Pythagoreans, Hippocrates thought that there was only one Comet, and it was considered to be a planet. He asserted that the Comet's tail was an optical illusion caused by the deflection towards the observer of sunlight in 'moisture' that the comet drew from the Sun whenever the Comet passed the Sun. Odd though it may seem, this explained why the Comet appeared to brighten and develop a long tail whenever it was in the Sun's vicinity. Hippocrates explained that whenever the Comet approached the Sun from the 'moister' northern skies it had no difficulty in drawing off enough moisture to produce a highly reflective tail, whereas if it was seen in 'drier' southern skies it did not appear to have a tail at all. He explained that the Comet's appearance lagged behind the annual revolution of the stars because its path gradually slowed down as it cleared the Sun.

Although these speculations were attempts to explain observed phenomena, it's easy to pick a number of gaping holes in them. How, for example, could the appearance of two comets at one time (or in very short succession) be explained? What might explain observations of those comets that attained peak brilliance and a large

Fig. 2.3 The cuneiform text on this ancient Babylonian clay tablet refers to observations of Halley's Comet that were made between September 22 and 28, 164 B.C.

tail far to the south of the Sun? The fact is that Hippocrates had access only to limited descriptive information about a few great comets that had been seen from his location in Mediterranean climes. Geometry dictates that great comets on high inclination orbits approaching from the north are more likely to be first spotted at night due to sky contrast; they develop lengthy tails as they head sunward, but as they move south their tails may appear to shorten because of the changing angle between Earth and the tail. When they are moving away from the Sun they may only be seen low to the horizon or in twilight conditions, where contrast reduces the brightness of the comet and the visibility of its tail (Fig. 2.4).

Although all original works of astronomer and mathematician Eudoxus of Cnidus (c. 410–347 B.C.) have been lost, we know that he authored a number of works, and it's possible to glean a great deal about his cosmological theories from secondary sources written around his time. Eudoxus made a number of significant contributions to astronomy. He introduced the celestial globe, built his own observatory in Cnidus and developed sophisticated models of planetary motion; he was also committed to the idea that the sphere and the circle remained perfect figures as befitted the motions of the cosmos.

Eudoxus' complete picture of the universe was geocentric, placing Earth at the center of a nest of 26 concentric crystalline spheres. To each of these spheres were

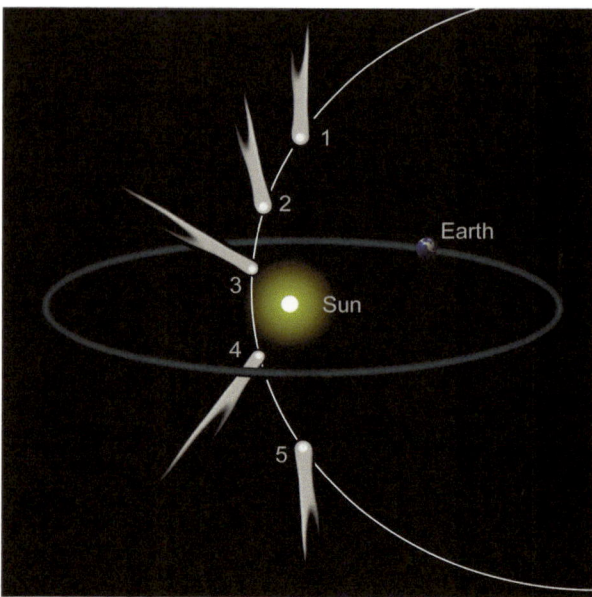

Fig. 2.4 Graphic showing how Hippocrates' theory of comets can be explained in terms of the apparition of a high-inclination comet, either moving from the north (Steps 1 through 5) or from the south (Steps 5 through 1). In the case of a comet moving north to south, geometry dictates (in this case) that the tail appears longest at Step 2, even though the tail is actually lengthier near perihelion (Steps 3 and 4, the latter not visible as it lies behind the Sun as seen from Earth). As the comet moves southward its tail is pointing away from both the Sun and Earth, and appears at its shortest (Illustration © by the author)

attached the Sun, Moon and planets (whose complex motions were determined by multiple associated spheres and whose axes of revolution and speeds differed), while the outermost sphere contained the distant stars that shared its axis with Earth and revolved once a year. It isn't known whether Eudoxus believed that this was the way things actually were, but the model seemed to explain the way that most celestial objects appeared to move and was the first full-blooded attempt to come to grips with the complexities of the observed motions of the heavens. One notable absence from Eudoxus' work—an absence doubtless explained by the fact that his ideas have only been partly preserved in the work of other philosophers—is the question of comets (Fig. 2.5).

Aristotle's Atmospherics

A century later, Aristotle (384–322 B.C.), one-time tutor to Alexander the Great (356–323 B.C.), took on Eudoxus' model of the cosmos and improved it by adding yet more spheres. His system contained no fewer than 55 of them. Aristotle

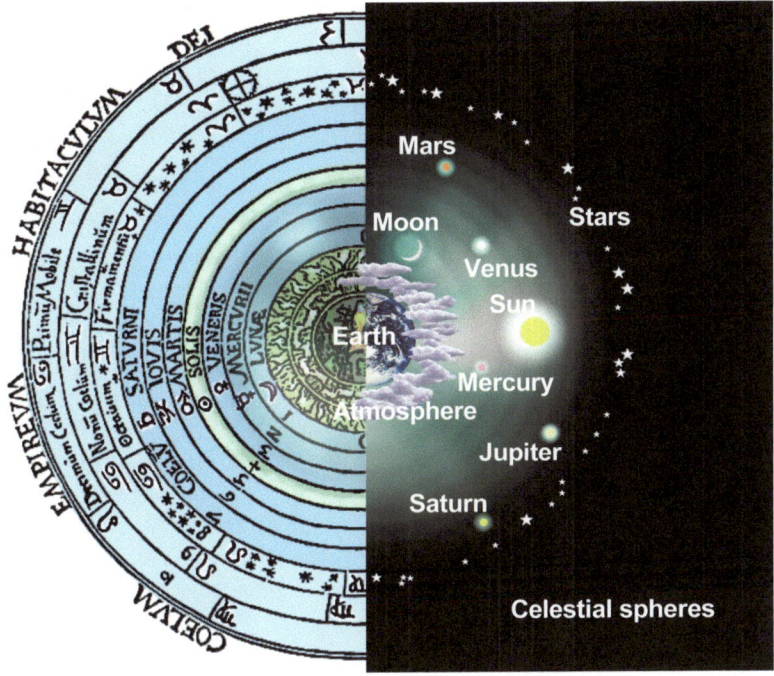

Fig. 2.5 Simplified diagram of the ancient Greek geocentric model of the Universe, based on illustration in Peter Apian's *Cosmographia* (1539) (Illustration © by the author)

considered that the heavens were perfect and unchanging. His cosmological model was that of a universe that was infinite in both past and future. There was only one place to allocate those errant, unpredictable comets. He ascribed them to phenomena that manifested in the upper atmosphere, above the clouds but below the perfect realm of the Sun, Moon, planets and stars.

Incidentally, we have Aristotle to thank for the very word 'comet.' It is derived from the Greek word *komē*, which means 'hair of the head.' Aristotle described comets as *komētēs* (meaning 'long hair'), an epithet that derives from the wispy appearance of a comet's tail. Anyone who has been fortunate to marvel at a great comet—one that is easily visible with the unaided eye at its peak brightness and displays a sizable coma and/or long tail—will know precisely why this description is particularly appropriate. To get an idea, just take a look at, say, an image of Comet McNaught (C/2006 P1) at its best; it looked like a vast plume of hair reefing in the wind, frozen in time (Fig. 2.6).

Aristotle made a concerted attempt to explain the nature of comets in the first book of his great work *Meteorology*, where he first explained, and then refuted, a number of speculations by other philosophers about the nature of comets. He rejected the idea that the Comet (singular) existed as an extraordinary manifestation of one of the five classical planets—Mercury, Venus, Mars, Jupiter or Saturn—because sometimes all

Fig. 2.6 The Great Comet of 2007 had a very 'hairy' tail structure. C/2006 P1 (McNaught), imaged by Robin Whittle as it appeared over Mt Macedon, northeast of Melbourne, Australia, on the evening of January 25, 2007. Sony DSC-F707, F2.2, 30 s

five planets plus the Comet could be accounted for. Occasional conjunctions and coalescences of two planets giving rise to a temporary comet, such as had been proposed by Anaxagoras (500–428 B.C.) and Democritus (c. 460–370 B.C.), were of course out of the question.

Aristotle declared that the idea of there being a single Comet was impossible because two comets have often appeared in the sky simultaneously. Hippocrates' theory of comets was subsequently demolished; such views involved certain impossibilities. He stated that if Hippocrates was correct then some bright comets ought to be visible before they had time to develop a tail, as they drew up reflective vapors as they neared the Sun; early on in their apparition they would appear as a bright point of light just like a planet. This was not so. As for comets only appearing in the north, Aristotle stated that many had actually first appeared in the south. In fact, comets could appear in any part of the sky, including directly opposite the Sun.

In particular Aristotle points to the famous apparition of a bright comet that arose in the west at the time of an earthquake and tidal wave at Achaea. Astronomers have since identified this as the Great Comet of 373–372 B.C., thought to have been one of the greatest comets in recorded history. It was an event that had profound consequences, because in all likelihood Aristotle himself saw this comet when aged 11, igniting his curiosity and perhaps inspiring him to follow the path of science and philosophy.

Heliocentric Anticipation

Although the model of an Earth-centered universe remained pre-eminent among thinkers in ancient Greece, it was by no means the only model, as we've already seen with Philolaus' fifth century B.C. theory of a Central Fire around which the Sun, Earth, Counter-Earth and all the planets revolved. In the third century B.C. the great mathematician and astronomer Aristarchus of Samos (310–230 B.C.) proposed a heliocentric model, prefiguring the work of Nicolaus Copernicus (1473–1543) by 1,800 years. Aristarchus proposed that the Sun lay at the center of the universe, and Earth (orbited by the Moon) was one of six planets that orbited the Sun; he even had the planets in their correct order from the Sun. The stars themselves were incredibly far away because they showed no parallax—in other words, the stars always appeared in the same place with regard to each other on the celestial sphere, no matter where in Earth's orbit we viewed them from. Aristarchus' views on comets are not known. Since he was bold enough to remove Earth from the center of the universe it would seem odd that he thought of comets as sporadic manifestations in the atmosphere; then again, comets would been impossible to accommodate in a heliocentric universe were they to have followed circular paths.

Despite the fact that Aristarchus' heliocentric universe was a far more elegant, streamlined and economic theory than any geocentric ones had been (or were to be proposed)—and one that explained some of the complex motions exhibited by the planets in an uncomplicated manner—it never took root. To think that the big, solid Earth was a planet in motion, revolving around both its own axis and the Sun, was to seemingly defy our experience. It was obvious that the heavens revolved around Earth because it looks that way. If Earth were spinning on its axis and zipping around the Sun, then surely, the argument went, we would feel this motion and observe manifestations of phenomena caused by it, such as high winds and no mean degree of dizziness. Besides, to remove Earth from the center of all things was to demote the status of our world and its inhabitants, ranking it alongside all the other objects in orbit around the Sun and retaining just the Moon for geocentric company. Clearly, the philosophical and religious consequences of the heliocentric universe were too unpalatable to bear, and remained so for a very long time indeed.

Ludwig Wittgenstein (1889–1951) took a famous view of this problem when he asked a friend: "Tell me, why do people always say that it was natural for men to assume that the Sun went around the Earth rather than the Earth was rotating?"

"Well, obviously," said his friend, "because it just looks as if the Sun is going around the Earth."

The philosopher replied: "Well, what would it look like if it had looked as if the Earth were rotating?"

Ptolemy's Long-Lived Legacy

Several centuries after Aristarchus, long after the classical period of ancient Greece had ended and the state had been subsumed into the Roman empire, there nevertheless appeared a number of important thinkers on the nature of the cosmos. Claudius Ptolemy (c. A.D. 90–168), of Greek lineage but a citizen of Alexandria in Egypt (a province of Rome), was perhaps the most important of these later philosophers. Ptolemy authored a number of scientific works. One of these, the *Almagest*, was an encyclopedia of ancient knowledge whose mathematical and astronomical concepts included an Earth-centered universe that attempted to explain the apparent motions of the heavens along the same sorts of lines as those propounded by Aristotle. *Almagest* is itself an Arabic word and derives from 'al-majisti,' referring to its ancient Greek title of *The Greatest Compilation*. Most of our knowledge of ancient Greek philosophy comes from original ancient texts (most of which have now been lost) that were translated, copied and preserved by Arab scholars in Baghdad during the European Dark Ages.

Ptolemy's model of the universe, expounded in *Planetary Hypotheses*, differed in one important aspect from that of Aristotle: the introduction of epicycles. Instead of each planetary motion being dependent on the individual speed and movement of several associated spheres, Ptolemy proposed that each planet described a small circular path around a point on a larger circular orbital path around Earth. Epicycles seemed to explain the observed looping movements of the planets—their so-called retrograde motions—while preserving the cosmologically sacred nature of the circle.

Contrary to his mathematical approach to the workings of the universe—and at odds with his generally logical way of thinking—it appears that Ptolemy made little attempt to explain what comets were and in which heavenly realm they appeared. Instead, in his work *Tetrabiblos*, his cometary speculations delved into the mystical and supernatural, attributing wars, storms and the fates of individuals to them. At the same time, however, he hints at their varied physical appearance, including their occasional resemblance to beams, trumpets and pipes, among other forms. He also indicated that some comets became visible near the Sun during the brief moments of darkness afforded by total solar eclipses. Such Sun-grazing comets have indeed been observed throughout history, most of the more recent ones having been discovered on images taken by the orbiting SOHO satellite (NASA's Solar and Heliospheric Observatory, operational since May 1996).

The works of Aristotle and Ptolemy were the most influential ever written, and for many centuries served as the main fonts of Western scientific knowledge.

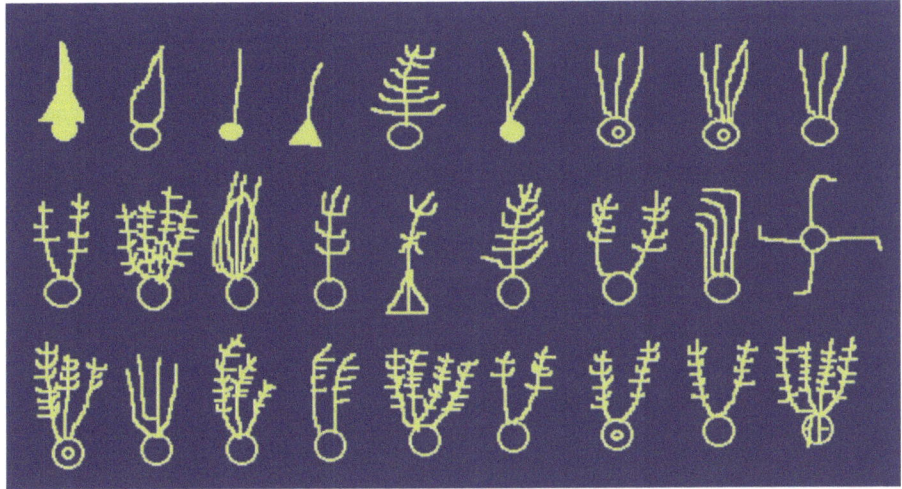

Fig. 2.7 In the Far East, comets—known as 'broom stars'—were regarded as omens, so it was considered vitally important to keep a close watch on celestial events. Thousands of astronomical phenomena are recorded in Chinese annals dating back many centuries. Over an almost continuous period spanning the sixteenth century B.C. to the end of the nineteenth century, Chinese court astronomers were appointed to observe and record changes in the heavens. This legacy of almost 3,500 years' worth of astronomy provides a rich source of reference material. Shown here is a copy of some of the cometary forms featured in the *Mawangdui Silk*, a textbook of cometary forms and the various disasters associated with them that was produced in the third century B.C.

Helped in no small way by assimilation into the dogma of the then-all-powerful Church, Aristotelian science and Ptolemy's epicyclic model of the workings of the universe came to assume an unchallengeable authority that did little to advance scientific reason, inquiry and progress until the Copernican revolution began in the early fifteenth century (Fig. 2.7).

Before leaving classical antiquity it's worth noting the remarkably near-the-truth opinions of the Roman playwright and philosopher Lucius Seneca (c. 4 B.C.–A.D. 65), mentor to Roman emperor Nero (A.D. 37–68). In his treatise *On Comets* Seneca declared the idea of there being but one Comet to be wrong, ruled out the suggestion that comets were illusions caused by planetary conjunctions and also dismissed the notion that comets were merely transient atmospheric occurrences. While taking the view that comets formed in the atmosphere by some sort of condensation of vapors and terrestrial 'exhalations' that were expelled into space, they still became real, solid, celestial objects with their own individual orbits. Comets could appear at any point in the sky, and they brightened as they approached the Sun. Although it may be possible to see distant stars through their tails, the heads of comets were opaque and therefore had real substance. He pointed out the great variety of curved celestial paths taken by comets and the varied shapes, sizes and levels of brightness that they assumed (he also mentions observer bias), but held that all comets were of the same nature. Following this logic, Seneca wrote: "Innumerable comets revolve in secret, unknown to us, either by the faintness of their light, or the situation of their orbit being such that they become visible only while they reach its extremities."

It may not come as a surprise to learn that there's no happy ending to the story of a great but complex man whose life was lived close to a paranoid and superstitious dictator (a trait shared by many dictators through history). Nero's reign saw the appearance of several bright comets, terrifying the emperor, who thought they were ill omens; Seneca attempted to make political capital by allaying such fears, assuring his leader that they were signs in Nero's favor. Sadly, the comet of A.D. 65 was to portend Seneca's demise. Suspecting that Seneca was plotting against him, Nero compelled him to take his own life.

There could be no more appropriate rounding-off to our brief survey of classical antiquity and the status of comets within its various cosmologies than to quote Seneca in *On Comets*:

> The day will yet come, when the progress of research through long ages will reveal to sight the mysteries of nature that are now concealed. A single lifetime, though it were wholly devoted to the study of the sky, does not suffice for the investigation of problems of such complexity. And then we never make a fair division of the few brief years of life as between study and vice. It must, therefore, require long successive ages to unfold all. The day will yet come when posterity will be amazed that we remained ignorant of things that will to them seem so plain. The five planets are constantly thrusting themselves on our notice; they meet us in all the different quarters of the sky with a positive challenge to our curiosity.

> The man will come one day who will explain in what regions the comets move, why they diverge so much from the other stars, what is their size and their nature.

> Many discoveries are reserved for the ages still to be when our memory shall have perished. The world is a poor affair if it does not contain matter for investigation for the whole world in every age…Nature does not reveal all her secrets at once. We imagine we are initiated in her mysteries. We are, as yet, but hanging around her outer courts.

New Observations Alter Old Theories

Copernican Revolution

Even though Aristarchus, in third century B.C. Greece, had proposed that the Sun, not Earth, lies at the center of the Solar System, the modern acceptance of the heliocentric theory begins with Nicolaus Copernicus (1473–1543). Copernicus, a mathematician and astronomer born in Poland but widely traveled around Europe, was inspired to inquire into the workings of the universe after learning about Ptolemy's theories. By combining brilliant insight with observations, he discovered that Ptolemy's theory of the Moon's motion was simply not good enough, planting

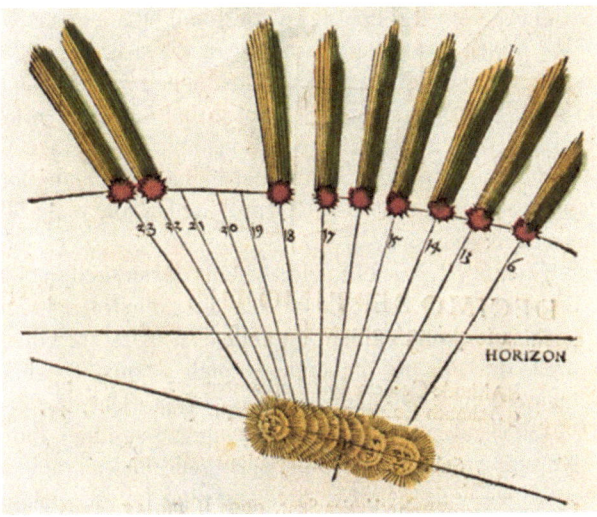

Fig. 2.8 An illustration from Peter Apian's book *Astronomicum Caesareum* (1540) depicting the comet of 1531 (now known as Halley's Comet); it shows that a comet's tail points away from the Sun (Courtesy of the Royal Astronomical Society)

seeds of doubt in his mind about the whole geocentric theory. Copernicus had access to many astronomical and mathematical works written in antiquity, and he doubtless became aware of Aristarchus' heliocentric theory.

Around 1510, Copernicus produced several handwritten copies of a manuscript, the *Little Commentary*, a work intended to be a private preparatory sketch (passed among friends) for a planned book detailing his proposed heliocentric theory. Although that planned book took many years to finally appear in print, Copernicus' heliocentric views were no great secret. In 1533 his theory was even heard by Pope Clement II and his cardinals in a lecture given in Rome by papal secretary Johann Widmannstetter. Indeed, Widmannstetter was so impressed that he later urged Copernicus to publish his theory, along with the data that backed it up.

Interestingly, the fourth decade of the sixteenth century also saw a remarkable succession of bright comets, much to the excitement of European astronomers who had endured a dearth of comets over the course of several preceding decades. Comets appeared in 1531, 1532, 1533, 1538 and 1539, drawing so much interest and attention that a flurry of tracts on comets appeared, written by contemporary astronomers such as Peter Apian (1495–1552), Gemma Frisius (1508–1555) and Girolamo Fracastoro (1478–1553) (Fig. 2.8).

Copernicus is sure to have marveled at the sight of these comets, and we know that he made a special study of one of them, the Great Comet of 1533, since he published a treatise on it. Incredibly, it appears that Copernicus failed to make the leap (at least in print) of realizing that comets were in orbit around the Sun. At the very least, he simply ignored the problem. In keeping with his contemporaries

Apian, Frisius and Fracastoro, Copernicus considered that comets were 'sublunary' phenomena of the upper atmosphere, just as Aristotle had thought eighteen centuries before, and the notion of the 'perfect' celestial path being bound to the form of the circle held firm. This may seem odd, given that Copernicus had made such a great leap in making Earth a mere planet. It is also puzzling, since a series of observational studies of comets, made a century before by the Italian mathematician and astronomer Paolo Toscanelli (1397–1482), provided ample material from which to arrive at the correct conclusion.

In the fifteenth century Toscanelli made what is considered to be the first modern attempt to study the paths of comets across the skies, his folios recording his observations of the apparition of a number of comets. The first, the comet of 1433, consisted of a simple drawing of the comet's path against the background stars, presented as a smooth curve without precise positional data. Later comets—those of 1449–1450, 1456 (Halley's Comet), the spring and summer comets of 1457 and that of 1472—were given a more thorough observational treatment in which careful positional observations with regard to stars and planets were made with the aid of naked-eye sighting devices, while timings were made using a clock.

Although these later studies show unevenness in the cometary paths due to slight observational errors in position, they are the more honest and scientifically valuable of Toscanelli's observations (indeed, they have remained useful to astronomers to the present day). Toscanelli's notes on his observations are, however, scant, and those which remain are mired in astrological babble. Although there is a hint in one of his diagrams that he attempted to measure the distance of comets by means of parallax, or at least understood the concept, there is nothing in his writings following this up; instead, he made little scientific use of his observations and fell back on the age-old default position of thinking that comets were high atmospheric phenomena (Fig. 2.9).

Given the evidence available to Copernicus it seems improbable that the idea that comets were in orbit around the Sun could have failed to enter his mind; we have no record of this, but an admission that their orbits were clearly not circular or centered on Earth may have presented him with an obstacle too great to take on in print. It was only towards the end of Copernicus' life, after having built up a great deal of knowledge and amassed considerable observational data, that Copernicus considered the time was right to publish his great work *On the Revolutions of the Heavenly Spheres* (1543).

On the Revolutions threw out the very fundamentals of Ptolemy's geocentric view of the universe. Copernicus stated that the stars are at an immense distance, compared to the distance from Earth to the Sun. He was convinced that the Sun, not Earth, lay near the center of the universe, and that the apparent daily rotation of the stars is caused by Earth's rotation. Copernicus went on to explain that the apparent annual circuit of the Sun around the ecliptic is caused by Earth revolving around the Sun, and the apparent retrograde motion of the planets is caused by the motion of Earth along an orbit inside that of the outer planets. His explanation of the phenomenon of retrograde motion dispensed with the need to introduce epicyclic planetary motions and is perhaps the most insightful and original of Copernicus' points.

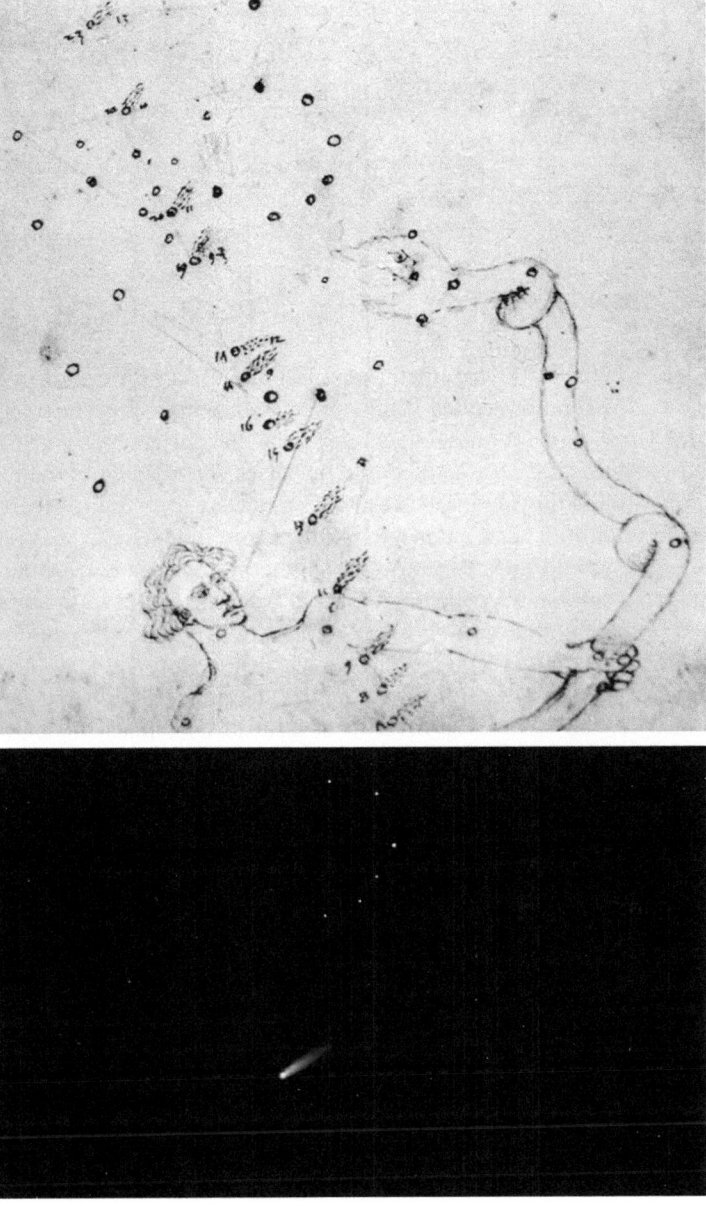

Fig. 2.9 Naked-eye observations of the comet of 1449–1450, shown moving from Ophiuchus through Corona Borealis, as depicted in the notebook of Paolo Toscanelli (*top*), and the author's simulation of the view on October 17, 1449

Comets are mentioned only once in *On the Revolutions*, where Copernicus wrote:

> It is said that the highest region of the air follows the celestial motion. This is demonstrated by those stars that suddenly appear. I mean those stars that the Greeks called comets. The highest region is considered their place of generation, and just like other stars they also rise and set. We can say that this part of the air is deprived of the terrestrial motion because of its great distance from the Earth.

Cosmic Conundrums

Tycho Brahe (1546–1601), regarded as the last and greatest astronomer of the pre-telescopic era, was a hot-headed Danish nobleman with a deep interest in mathematics and astronomy. Although he admired the elegant geometry of Copernicus' heliocentric theory the fact that he could not discern any stellar parallax owing to Earth's orbit around the Sun implied that the stars were at an unimaginable distance from us. He found this idea untenable. At the same time, Tycho was unwilling to dispense entirely with the geocentric Ptolemaic system because of the religious-philosophical connotations of removing Earth from the center of the universe.

Tycho set about constructing his own scheme of the universe, a geo-heliocentric model known as the Tychonic system, in which the Sun and stars orbited Earth while the other planets orbited the Sun. In this scheme, Earth remained at the center of the universe; the stars were fixed to a vast, all-encompassing Earth-centered globe, removing the problem of stellar parallax.

As early as 1563, when aged 17, Tycho had been painfully aware of the inaccuracy of current maps of the heavens, writing:

> I've studied all available charts of the planets and stars and none of them match the others. There are just as many measurements and methods as there are astronomers and all of them disagree. What's needed is a long term project with the aim of mapping the heavens conducted from a single location over a period of several years.

Realizing the need for more accurate positional data about the movements of the planets to get to the truth, Tycho began to make precise measurements with the aid of naked-eye quadrants and cross-staffs. However, Tycho's careful astronomical observations went on to provide plenty of evidence *against* age-old Aristotelean notions of an Earth-centered universe (Fig. 2.10).

Purely by chance, a vital piece of evidence against Aristotle's picture of 'unchanging heavens' came in the form of a brilliant 'new star' that suddenly appeared in the constellation of Cassiopeia in November 1572. Now known as SN 1572, the event was a Type Ia supernova—the catastrophic explosion of a white dwarf star—which, after initially rivaling Venus in brilliance, gradually faded but remained visible to the unaided eye into 1574. SN 1572 was one of the most important single astronomical phenomena in history, because careful positional measurements by Tycho and his contemporaries showed that it displayed no discernible parallax; its observed position in the skies when viewed from different locations and times remained the same.

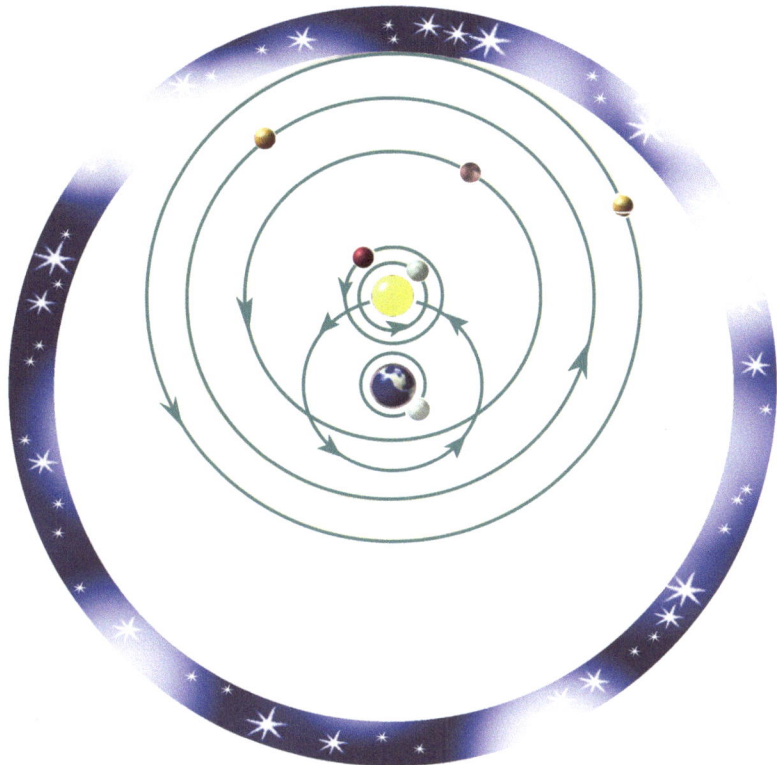

Fig. 2.10 A depiction of the Tychonic geocentric system (not to scale). Both the Sun and Moon revolve around Earth, while Mercury, Venus, Mars, Jupiter and Saturn orbit the Sun. In this manner, neither Mercury or Venus appear to stray very far from the Sun, while the outer planets Mars, Jupiter and Saturn regularly appear at opposition (opposite to the Sun in the sky). Comets posed a problem—they were clearly located in planetary realms and didn't keep to circular orbits around either the Sun or Earth. The entire Solar System, in this scheme, is surrounded by a sphere of fixed stars (Illustration © by the author)

No conclusion other than that the new 'star' lay in distant starry realms was possible; but it was a conclusion that struck at the fundamental Aristotelian notion of eternal celestial immutability. In 1573 Tycho published his observations of the supernova in *Concerning the New Star*, metaphorically striking the first nails into Aristotle's cosmic coffin. Tycho's account of the phenomenon—the first scientifically studied supernova—was published in Copenhagen in 1573 and was the basis for his subsequent reputation as a first-class astronomer (Fig. 2.11).

Aware of Tycho's importance in upholding the status of Denmark as a progressive nation that nurtured the advancement of science, King Frederick II granted Tycho a large estate on the island of Hven in Copenhagen Sound (the island now belongs to Sweden), along with generous funding to establish an observatory there. Before long, the astronomer had set about creating the most modern observatory of

Fig. 2.11 Tycho's observation of the supernova of 1572, compared with a simulation of what it would have looked like in the sky (Illustration © by the author)

the day. Between 1576 and 1590, a large castle-styled observatory known as Uraniborg (from its dedication to Urania, the muse of astronomy) grew at the island's center. Astronomical observations were made using a variety of skillfully fabricated instruments, including a mural quadrant, revolving wooden and steel quadrants, astronomical sextants and equatorial armillary spheres, many of them featuring novel designs of Tycho's invention (Fig. 2.12).

From Uraniborg, Tycho's meticulous naked-eye observations were to strike yet more nails into Aristotelian and Ptolemaic cosmology. He kept a meticulous 10-week observational record—some 24 measurements made between November 1577 and January 1578—of the track of a comet that he had first sighted in November 1577, carefully following it as it traversed the northern skies. Ten years later he published his work on the comet (now designated C/1577 V1) in *Concerning New Phenomena in the Ethereal World*, in which he pointed out that his observations of the comet's tail disproved Aristotelian notions:

> [A]t all times this comet had its tail turned directly away from the sun, as all other comets, those observed many years ago by Regiomontanus, Apian, Gemma Frisius, and Fracastoro, have also done: all have turned their tails away from the Sun. From this, it appears that the tail of a comet is nothing but the rays of the Sun which have passed through the body of the comet…Therefore, Aristotle and all those who follow him cannot maintain their opinion, namely that the tail of a comet is a flame of the rare fattiness which is burning above the air, for if that were true, these flames would not have a relationship to the Sun, and always turn themselves away from it.

Far from being an object high in Earth's atmosphere, Tycho was certain that the comet lay much further away than the Moon because simultaneous observations made with a confederate in Prague, 660 km south of Uraniborg, showed that it displayed a great deal less parallax than the Moon. He went on to plot the comet's path, placing part of its orbit near that of Venus. Tycho suggested that cometary

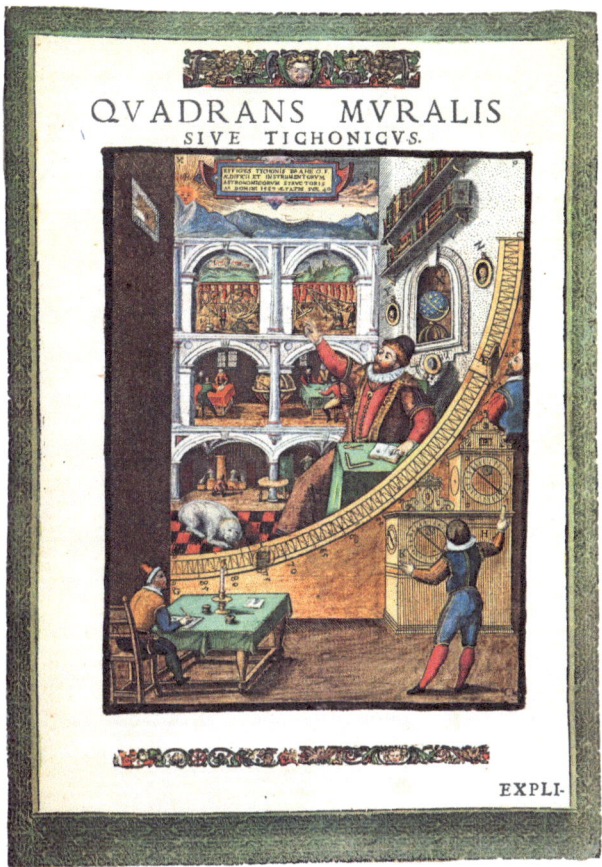

Fig. 2.12 Tycho Brahe, pictured at work using a large mural quadrant in his observatory on the island of Hven. From Tycho's Mechanical Astronomical Instruments (1598)

orbits may not keep to circular paths and also remarked on the fact that the comet's tail always pointed in a direction away from the Sun, regardless of the comet's motion through space. There wasn't any doubt that this comet was located in true planetary realms, a place where Aristotle had claimed that "nothing new could be born." Tycho's 400-year-plus observations of C/1577 V1 have since been used to pinpoint its likely current location, some 300 au from the Sun, far above the plane of the Solar System.

Yet all was not tempered by hard data, mathematics and reason; despite finding evidence to support the idea that Earth, along with planets and comets, were in orbit around the Sun, Tycho remained highly skeptical of the idea. He clung to the old notion of an Earth-centered universe whose motions would eventually be explained as soon as the right mathematical model was found. Added to his scientific insight,

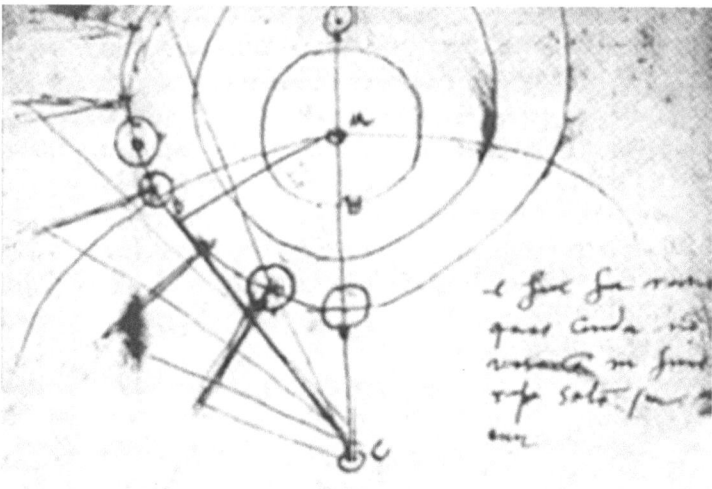

Fig. 2.13 The path of the comet of 1577, from Tycho's notebooks

Tycho, like so many of the era, attached a useless layer of astrological significance to comets. However, these startling discoveries were among the first of what was to become a flurry of thin ends of wedges that were to be driven into the once-solid edifice of dogmatic belief in ancient cosmological ideas during the sixteenth and seventeenth centuries (Fig. 2.13).

Towards a Deeper Understanding: The Laws of Planetary Motion

As a young boy Johannes Kepler (1571–1630) had marveled at the sight of the great comet of 1577, an object that had proven so difficult for Tycho Brahe to admit into his scheme of the universe. However, his curiosity about the skies was never destined to be realized in as practical a manner as that of Tycho, since his ability to perform the tasks of observational astronomy was impaired by limb weakness and poor eyesight consequent to childhood smallpox. It is a little-known fact, however, that Kepler used a telescope in his later years, and in September 1618, he became the first person to have ever observed a comet telescopically—comet C/1618 Q1 (see below).

    Instead, Kepler's genius lay in the use of mathematics to achieve a firm grasp on the movements of the heavens and the motions of the Moon and planets. Profoundly religious, he was convinced that the universe had been created in accordance with certain mathematical rules, and that knowledge of these rules was within human comprehension. His work inspired yet greater understanding among later generations of astronomers, and indeed his legacy remains potent to this day.

Perhaps the great comet of his childhood had imbued within Kepler a deep sense of the transcendent and mystical, since, in addition to astronomy, he was a firm believer in astrology—a subject that we now consider to be the unscientific art of attempting to reconcile planetary movements with Earthly phenomena. It is strange to think that any rational person could give any credence to astrology, but in Kepler's time the lines between scientific logic and ill-founded superstition were somewhat blurred, particularly when it came to the subjects of astrology and astronomy. It was actually through the practice of astrology as a student at the University of Tübingen in Germany that Kepler first developed an understanding of the apparent motions of the planets. In later life, while at Prague, Kepler's chief role as imperial mathematician to Rudolph II was to provide astrological advice.

In October 1604—just 32 years after Tycho's supernova—another 'new star' flared into being. This time, the supernova occurred in the constellation of Ophiuchus and peaked at around the brightness of Venus. Kepler thought that the supernova affirmed the fact that the heavens could no longer be regarded as change-less and immutable. The new star was undoubtedly in stellar realms because, like other stars, it displayed no measurable parallax.

Kepler also interpreted the 1604 supernova astrologically, believing that it represented the beginnings of an important new phase of terrestrial events. Although his astrological predictions and horoscopes may have possessed little scientific merit (a quality shared by all astrological forecasts), Kepler's painstaking analysis of Tycho's observations was to produce one of the most important scientific insights of all time. Tycho had originally directed Kepler to investigate Mars' orbit using a mathematical tool known as an equant, helping solve some of the problems observed in planetary motions introduced by the geo-heliocentric Tychonic system. Kepler eventually created a model that agreed with observations to a point, but still produced discrepancies between observation and theory of up to 8 arcminutes (almost one-quarter the Moon's apparent diameter) (Fig. 2.14).

In 1600 Kepler visited Tycho in order to analyze the great astronomer's meticulous observations—measurements of the motions of Mars in particular—in order to discover whether a set of fundamental laws about planetary motions could be determined. Using Tycho's extensive records—the most accurate and consistent set of naked-eye measurements ever made—he was to refine Copernicus' heliocentric theory and place it on a firm scientific footing.

After analyzing Tycho's observations of Mars from 1587 to 1595 he dispensed with perfect circles and equants and went to work on the basis of the idea that Mars traveled an ovoid (egg-shaped) path around the Sun. He discovered that the planets move faster when nearest the Sun, and the radius vector (the line connecting a planet to the Sun) sweeps out equal areas in equal times, a concept now known as Kepler's second law of planetary motion.

Kepler's next discovery came with his insight into the shape of Mars' orbit when he hit upon the idea of an ellipse, a shape formed by cutting through a cone at an oblique angle. He realized that for years he had been attempting to fit an ovoid into an oval slot. It made sense that all planetary orbits must be ellipses of varying degrees—that of Mars being particularly eccentric among the planets—and the Sun is located at one focus of this ellipse. This is known as Kepler's first law of planetary

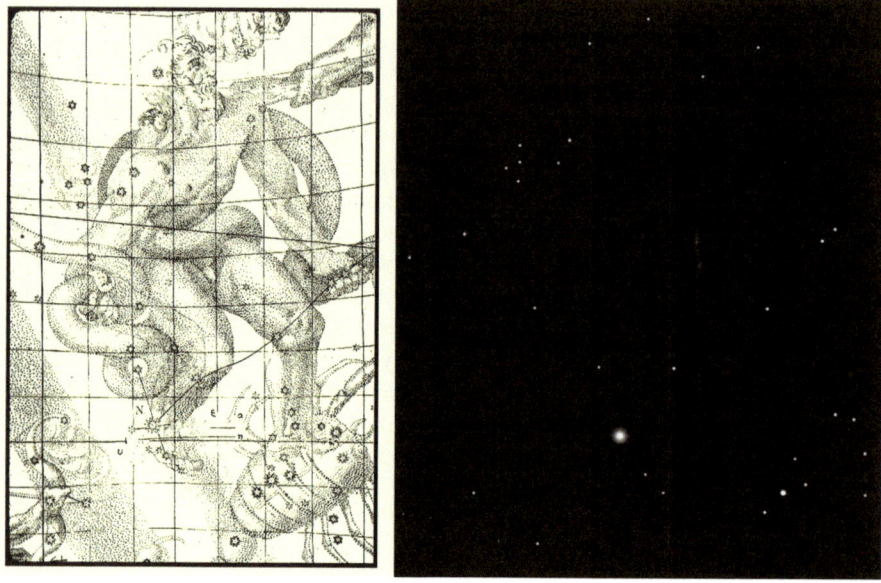

Fig. 2.14 Kepler's published observation of the supernova of 1604, compared with a simulation of how it probably appeared in the sky (Illustration © by the author)

motion. Describing this 'eureka' moment, Kepler wrote: "I awoke as from a sleep, and new light broke upon me."

Yet, even in Kepler's new scheme, this 'new light' was yet to fall upon comets. After studying the path of a bright comet that appeared in 1607 (which, unbeknown to him was the periodic visitor Halley's Comet), he came to the curious conclusion that comets moved freely through the Solar System, more or less in straight lines. However, it appears that the observational data available to Kepler were not good enough for him to compute an orbit accurate enough to betray the comet's curved path. Kepler's astrological inclinations caused him to believe that comets were omens, guided through space by an intelligent, often seemingly malevolent supernatural spirit; they influenced terrestrial events by disrupting nature itself.

Nevertheless, Kepler did come to some reasonable conclusions about comets, their physical properties and phenomena. He speculated that comets were spontaneously created from impurities or 'fatty globules,' in the ether; they were as numerous as fish in the sea, but we see only a small selection of them during our short time on Earth. Explaining cometary tails, he wrote:

> The direct rays of the Sun strike upon it [the comet], penetrate its substance, draw away with them a portion of this matter, and issue thence to form the track of light we call the tail…In this manner the comet is consumed by breathing out its own tail…the head is like a conglobulate nebula and somewhat transparent; the train or beard is an effluvium from the head, expelled through the rays of the Sun into the opposed zone and in its continued effusion the head is finally exhausted and consumed so that the tail represents the death of the head.

Four years after their discovery, the first two laws of planetary motion featured in his book *New Astronomy* (1609). The book also contains ideas about gravity (many decades before Isaac Newton's theory of gravitation) and speculation that the Sun's own position in space was far from being stationary. It wasn't until 1619 that Kepler's third law of planetary motion appeared in his book *Harmonies of the World*; it states that the ratio of the length of the semi-major axis of each planet's orbit (cubed), to the time of its orbital period (squared), is identical for all planets. Known as the 'harmonic law' this is perhaps better expressed by stating that the square of a planet's orbital period is proportional to the cube of its mean distance from the Sun.

Kepler's three laws of planetary motion could be applied to any object orbiting any other object under the influence of gravity. The laws appeared to explain the motions of all objects in the Solar System—including comets—and they delivered a means by which the scale of the Solar System could be deduced. Kepler's laws enabled Isaac Newton to lay out his theory of gravitation in his *Principia* (1687), which demonstrates that Keplerian orbits are the most simple of two-body orbits.

Chapter 3

The Telescopic Era

Comets Under Closer Scrutiny

Having heard about a newly invented optical instrument that made distant objects appear larger, Galileo Galilei (1564–1642), a professor of mathematics at Padua University, ground and polished his own lenses to make small, simple telescopes of his own. Although a number of people had made and used telescopes in the months before Galileo placed his own eye to the eyepiece, Galileo is acknowledged by most to have been the first to turn the telescope to the skies to make scientific observations.

If Galileo is to be considered the father of modern astronomy, he effectively disowned comets. Although he saw a number of comets with the unaided eye, he is not known to have observed any comet through the telescope. Although Galileo was a great proponent of the Sun-centered universe—many of his telescopic observations providing ample evidence for this—his views on the nature of comets were far from progressive. Despite the great advances made by Tycho and Kepler, both of whom thought that comets were real objects traveling through the Solar System, there lingered Aristotelian views, even in the minds of great scientists such as Galileo, that comets were simply manifestations of phenomena visible in Earth's atmosphere. He reasoned that comets were not burning objects, because if they did then they would have an effect on Earth's weather. No, Galileo considered comets to be caused by the atmospheric refraction of sunbeams in vapors rising from the ground.

The year 1618 saw the appearance of three bright comets in quick succession. Following comet C/1618 Q1 in August-September of that year (telescopically observed by Kepler) came C/1618 V1 in November-December, whose heels were

P. Grego, *Blazing a Ghostly Trail: ISON and Great Comets of the Past and Future*, 61
The Patrick Moore Practical Astronomy Series, DOI 10.1007/978-3-319-01775-4_3,
© Springer International Publishing Switzerland 2014

in turn hard-followed by C/1618 W1, visible from November 1618 to January 1619. The comets inspired a flurry of publications around Europe, including *On the three comets of 1618* (1619) by Orazio Grassi (1583–1654), professor of mathematics at the Roman College, who concluded that comets were real astronomical bodies in planetary realms, in orbit around the Sun. Grassi, a Jesuit who supported the helio-centric theory of the universe and accepted the telescopic discoveries of Galileo, must have been surprised when his ideas were rebutted as being 'false and vain' by Galileo himself in *Discourse on Comets* (1619) which was written by Galileo's student Mario Guiducci (1584–1646).

Grassi's response, leveled directly at Galileo, was delivered in his *Book of Astronomy and Philosophy* (1619); this was again countered by Galileo in his book *The Assayer* (1623), where he took a contrarian stance that harkened back to Aristotle, writing: "A comet is not one of the wandering stars which become visible in a manner similar to that of some planet." Galileo's battle was not forgotten by the Jesuits when, in later years, he went to trial for heresy, but Grassi himself remained distant from these bitter proceedings and their repercussions.

Despite the fact that progress in cometary thought was to some extent impeded by the opinions expressed by Galileo, a number of eminent scientists pursued origi-nal lines of thought. In the Netherlands Willebrord Snel (1580–1626), a mathemati-cian with a great interest in astronomy, was also inspired by comet C/1618 W1. In his *Comet Descriptions* (1619) he presented his own observations, along with notes on comet C/1585 T1 (which he may have seen as a young boy). Snel was strongly critical of Aristotle's views on comets and the unquestioning adherence to estab-lished authority. Yet paradoxically, unconvinced by the heliocentric system of Copernicus, he remained a firm believer in a geocentric universe.

As telescopes improved and the Enlightenment advanced, an increasing number of astronomers were keen to question and learn more about the universe by scrutiniz-ing celestial objects and their associated phenomena more closely than ever before. It was an age of new ideas, some healthy and productive, others not so plausible.

René Descartes (1596–1650) favored the idea of a mechanistic universe that behaved like a clockwork orrery, the motion of each of its components being able to be predicted. In *Principles of Philosophy* (1644) he put forward the principal laws of nature. His first law postulated that in the absence of external forces, each thing always remains in the same state; consequently, when it is once moved, it always continues to move. His second law prefigures Newton's first law of motion (see below): all movement is, of itself, along straight lines.

Descartes speculated that comets were faded old stars that had been 'slingshot' towards the Sun after having been passed from pillar to post by the 'vortices' of more energetic stars. Age-old superstitions about the *meaning* of comets contin-ued to thrive throughout the seventeenth, eighteenth and nineteenth centuries, including the widespread idea that they were sent by God as handy hints to humanity. Counterproductive and often harmful, irrational beliefs were aug-mented by (some even springing from) new scientific discoveries; we are familiar with such cometary confabulations to this day, since the twentieth and twenty-first centuries have, sadly, not been immune to nonsensical and pseudoscientific proclamations about comets (Fig. 3.1).

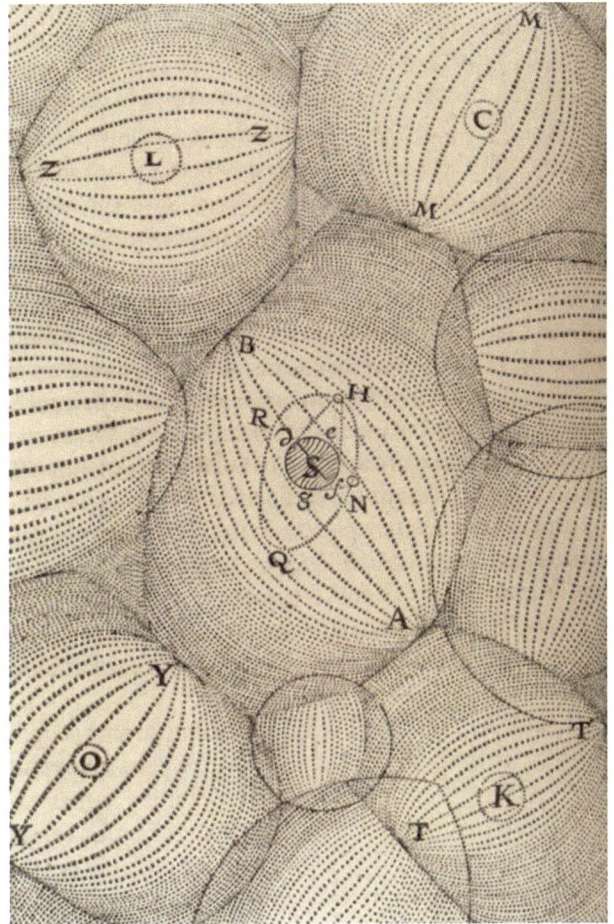

Fig. 3.1 An illustration from Descartes' *Principles of Philosophy* showing cosmic 'vortices'

In London, Francis Bacon (1561–1626) pondered the ways in which comets might be studied in order that their influence upon Earth become better understood, bringing new insights into real physical processes at work in the universe. Under the impression that droughts often followed the appearance of comets, he asked how could they produce such an effect if their blazing light produced no detectable warming like the Sun? To solve such problems, Bacon advocated a more thorough study of both the history of cometary apparitions; a more careful analysis of comets would lead to a 'sane astrology'; and the ability to predict comets in the future. In *New Organum* (1620) he wrote:

> Comets…have…power and effect over the gross and mass of things; but they are rather gazed upon, and waited upon in their journey, than wisely observed in their effects; spe-
> cially in their respective effects; that is, what kind of comet, for magnitude, colour, version of the beams, placing in the region of heaven, or lasting, produceth what kind of effects.

Meanwhile, in Paris, the great observational astronomer Pierre Gassendi (1529–1655) literally put his neck on the line in publishing *Dissertations against Aristotle* (1624) in full knowledge of the fact that Parliament had shortly before decreed that the death penalty would befall '[any] person should hold or teach any doctrine opposed to Aristotle.' Gassendi, a proto-materialist, rightly concluded that the small parallax of the comets of 1618 meant that they were at a great distance from Earth, just as Tycho had concluded. However, in order to avoid ecclesiastical wrath he changed his expressed opinion of Tycho's system of the universe, from skepticism to full support.

From his private, well-equipped observatory 'Star Castle' in Danzig, Poland, the wealthy brewer Johannes Hevelius (1611–1687) made a wide variety of astronomical observations. He is most famous for his lunar observations, resulting in the publication of *Selenographia* (1647), the most accurate Moon map of its day, and for his measurements of star positions that enabled the publication of the splendid *Uranographia* (1687), the most advanced celestial atlas of its time (accurate star maps were, and still are, essential in plotting the paths of comets). *Uranographia* contained charts of 73 constellations, including 12 introduced by Hevelius himself, of which the latter 7 names are still in use today, namely Canes Venatici (the Hunting Dogs), Lacerta (the Lizard), Leo Minor (the Lesser Lion), Lynx (the Lynx), Scutum (the Shield), Sextans (the Sextant) and Vulpecula (the Fox).

Hevelius discovered four comets — in 1652, 1661, 1672 and 1677 — all with his keen unaided eye. It is now thought that the second of his discoveries was the previous apparition of the brilliant comet 153P (Ikeya-Zhang), which bowled over its viewers on its return in March and April 2002 (see Chap. 4). Appropriately enough, in 1668 — slap between the publication of *Selenographia* and *Uranographia* — Hevelius produced the richly illustrated *Cometographia* (1668), in which he speculated that comets revolve around the Sun in parabolic paths (Figs. 3.2, 3.3, and 3.4).

Hevelius' contemporary, the pioneering Dutch physicist and astronomer Christian Huygens (1629–1695), viewed comets to be transitory objects that were just passing through the Solar System with a uniform motion through space. He observed several comets, including those of 1664 and 1680, and spent some time pondering how their paths could be reconciled with either Descartes' 'vortices' or Newtonian gravity (see below). Descartes' scenario required celestial objects to be moved along by some form of 'subtle matter,' 'fluid medium' or 'ether,' while gravity involved a seemingly inexplicable action at a distance by objects upon one another. Huygens realized that if any real physical substance filled space, however tenuous, then it would exert a resistance upon objects such as planets and comets, hence slowing them down over time and altering their course; nothing like that was observed.

Huygens' accurate telescopic observations (featuring the use of pendulum clocks for timing events) — along with cometary observations by a host of other European astronomers — were used extensively in what is the first true compendium of comets along the lines of that suggested by Bacon more than half a century earlier. *The Theater of Comets* (1668) by Polish astronomer and historian Stanislaw Lubieniecki (1623–1675) gathers accounts of cometary apparitions through history. One of the delights of this book is its range of engraved illustrations that were taken

Fig. 3.2 The title page of Hevelius' *Cometographia* (1668) shows three allegorical figures representing different ideas about comets. *At left* an Aristotelian postulates that comets are sublunar, while *at right* is a Keplerian maintaining that comets move in straight lines. The Hevelian, *at center*, thinks that comets originate in the atmosphere of Jupiter and Saturn and orbit the Sun on curved paths

directly from the observers' observations and star charts, giving a rich mix of styles of depiction, from heavily illustrated animistic maps of the heavens to neat, sparse and easily deciphered charts showing various cometary paths (Figs. 3.5 and 3.6).

On the Shoulders of Giants

From an early age Isaac Newton (1642–1726) had a great desire to understand the physical world. He was, of course, to become one of the greatest scientists in history, a true genius whose works made the universe understandable through his

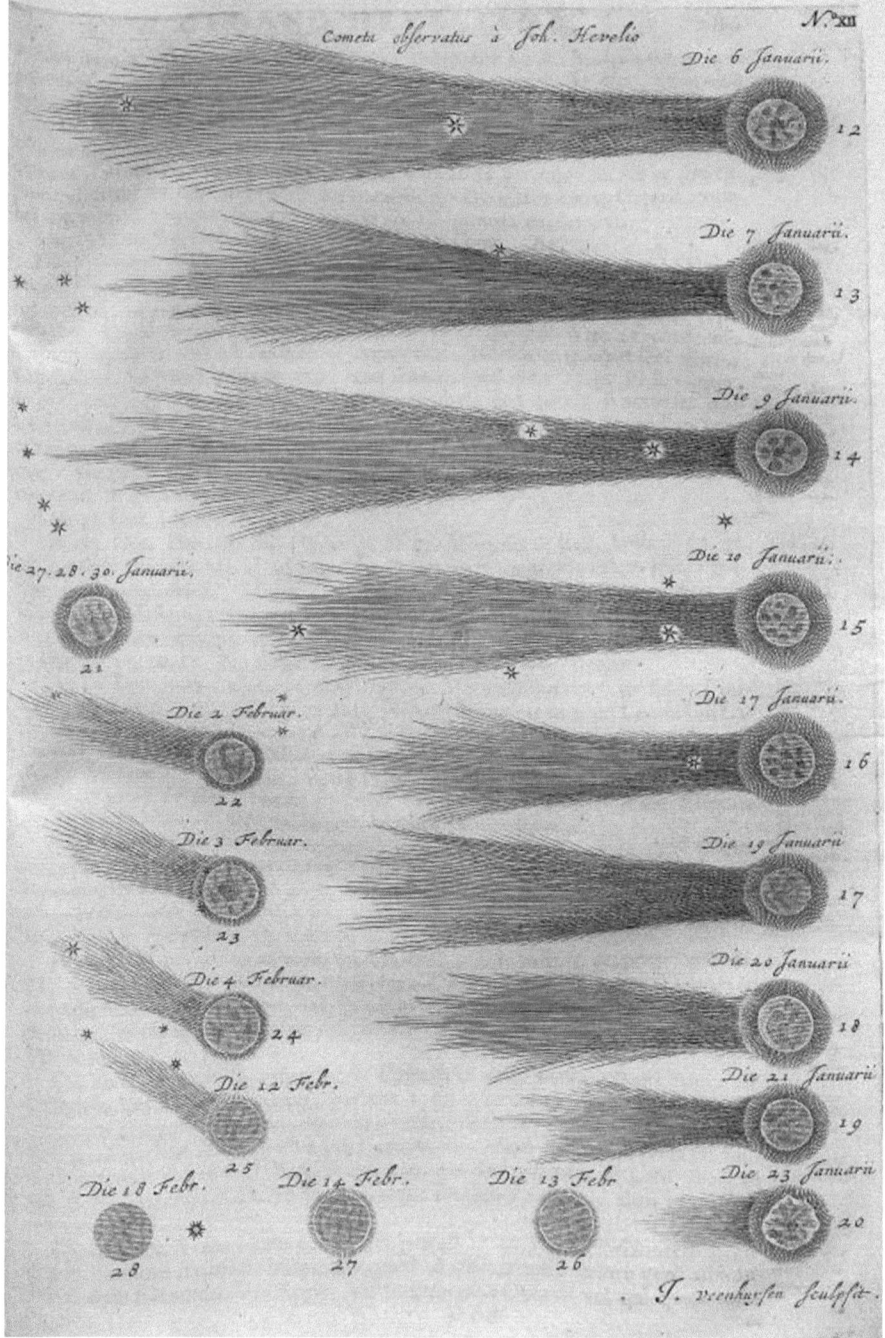

Fig. 3.3 Drawings of the changing appearance of the comet of 1665, from Hevelius' *Cometographia* (1668)

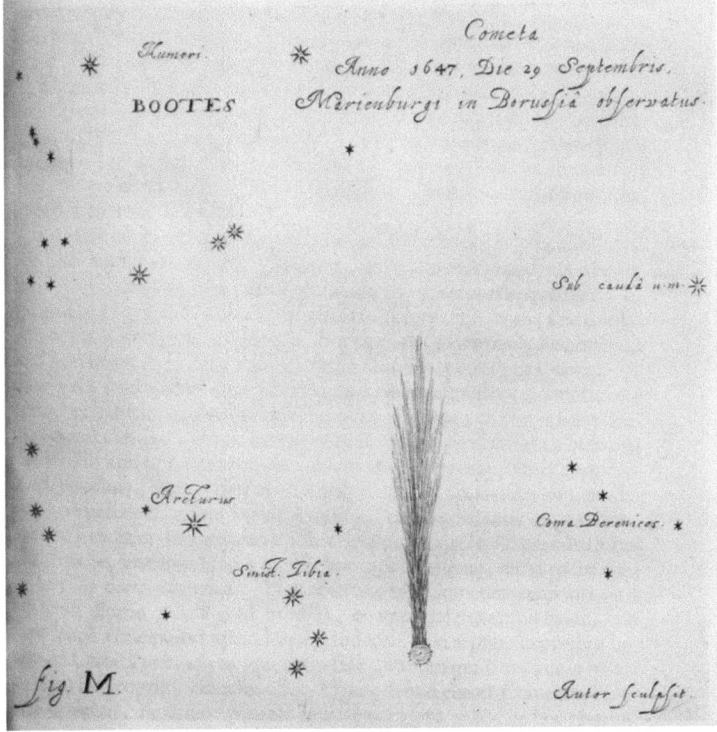

Fig. 3.4 Hevelius' observation of a comet, made on September 29, 1647, from Hevelius' *Cometographia* (1668)

universal laws, insights and inventions. Bringing together the ideas of great scientists preceding him, such as Copernicus, Galileo, Kepler and Descartes, Newton is famously quoted as saying: "If I have been able to see further it was only because I have stood on the shoulders of giants."

It is no exaggeration to state that Newton's incredible insights ushered in an entirely new era in our understanding of, and our capability to understand, the universe. His greatest period of work took place between 1665 and 1687, during one of history's most exciting times of scientific revolution. He developed the binomial theorem and calculus in mathematics, formulated his laws of motion and his law of universal gravitation in physics, while in optics he developed our understanding of light and invented the reflecting telescope—the basis of many of the world's largest telescopes. Newtonian mechanics dominated physics for more than 200 years, and although Einstein's general theory of relativity (1915) has transcended Newtonian gravity, it was Newton's laws that helped direct the space probe *Giotto* to Halley's Comet in 1986 (to give but one of many space age examples).

Newton's scientific interests were wide and varied, and it's not surprising to find that he turned his attention to comets at times. In his younger days he transcribed

Fig. 3.5 Part of an illustration of a cometary path through Orion and Taurus in 1666. From Lubieniecki's *The Theater of Comets* (1668)

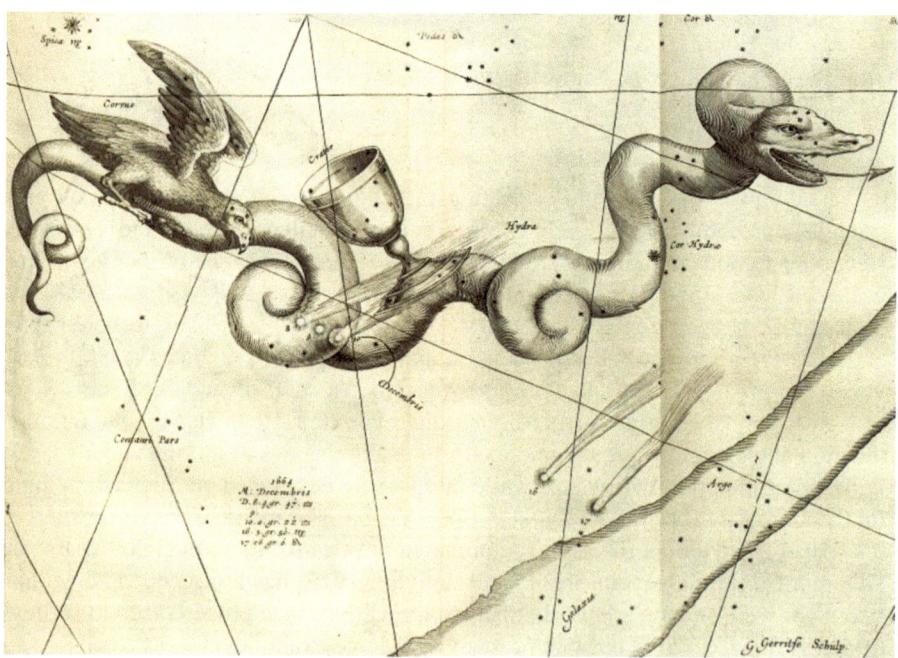

Fig. 3.6 Part of an illustration of a cometary path through Hydra, observed by Gaspar Schott in December 1664. From Lubieniecki's *The Theater of Comets* (1668)

and studied Snel's tables of the motion of the comets of 1585 and 1618. Of particular interest are his observations of comet C/1664 W1, which he followed between December 23,1664, and January 23, 1665. His notes feature a diagram showing the comet's position among the stars, but his descriptions and attempts to track its motion with references to the Moon's position showed a certain degree of naiveté; this is not surprising, given that it was the first time that Newton had seen a comet, and his excitement at viewing "a Comet whose rays were round her, yet her tayle extended it selfe a little towards east" is evident. The same comet is thought to be the one referenced by John Milton (1608–1674) in *Paradise Lost* (Book II, 1667):

like a comet burned,
That fires the length of Ophiuchus huge
In the artick sky, and from his horrid hair
Shakes pestilence and war

Far from being an ill-omen, C/1664 W1 caused Newton to think about how comets fit into the scheme of things and why it was that a general consensus of contemporary notions of comets and their paths through space was nowhere near to being settled on. But Newton's active mind soon moved on to other things, and as we shall see wasn't to return to comets in a major way until 1680.

Newton Solves a Weighty Problem

The Great Comet of 1680 is thought to have been significant in prompting Newton to test his burgeoning ideas on gravity. This was the first comet to have been discovered telescopically and was found by Gottfried Kirch (1639–1710) on November 14, 1680. It was extensively followed by Giovanni Cassini (1625–1712) at the Paris Observatory and John Flamsteed (1646–1719) at the Royal Observatory, Greenwich, for 4 months continuously. Now known as C/1680 V1, the comet traced a path across northern skies and brightened enough to be seen during the morning skies as it headed towards an extremely close flyby of the Sun. Once it had rounded the Sun in what was a hairpin turn the comet emerged from the solar glare as a brilliant evening object sporting a very long tail.

Despite having had such a close brush with the Sun, the comet's survival as a single entity caused Newton to think that the heads of comets were "solid, compact, fixed, and durable, like the bodies of the planets." He also speculated on the nature of cometary tails. On arriving in the Sun's neighborhood of the Sun, comets "… shine with a greater lustre, and send forth conspicuous tails, which are nothing but subtle vapors exhaled from the body of the comet by the intensity of the Sun's heat, as steam from water set over the fire…the most refulgent of these tails hardly appears brighter than a beam of the Sun's light transmitted into a dark room thro' a hole of a single inch diameter, and that the smallest fix'd stars are visible thro' them without any sensible diminution of their lustre."

It was obvious that the comets of late 1680 and early 1681 were just one comet—not the fortuitous apparition of two comets, one heading towards the Sun, the other heading away from it—and astronomers realized that its path must have been increasingly bent as it experienced its near-solar encounter. After analyzing his own observations, astronomer Georg Dörfel (1643–1688) produced a graphic showing that the comet's path through space was parabolic (an open-ended curve with the Sun at its focus) moving in a manner consistent with Kepler's laws of planetary motion.

Mathematician Jacob Bernoulli (1655–1705) also considered cometary paths could be calculated so that their past, present and future positions could be worked out accurately. Like Gassendi and Bayle, Bernoulli rejected the astrological significance of comets; in his treatise he labeled practitioners of astrology as 'shufflers and cheats.' Bernoulli was so confident of being able to predict cometary paths that he announced that C/1680 V1 would make a return in May 1719. Sadly, Bernoulli had neither the exacting observational data nor the advantage of a theory of gravity (then in the mind of Newton alone) to arrive at an accurate prediction, and his work incurred no mean degree of ridicule, especially when the comet failed to reappear. Bernoulli's prediction was even dredged up and ridiculed by the great philosopher Voltaire (François-Marie Arouet, 1694–1778) a century later in *Philosophical Letters* (1778). In that publication, Voltaire also dismissed Descartes' ideas about cosmic vortices, pointing out that, if they existed, they did not obey Newton's laws (see below).

Hevelius, Dörfel and Bernoulli, along with their contemporaries, were completely at a loss to explain how comets, let alone the planets, were able to sweep through space along certain well-defined curves, be they hyperbolas, parabolas, ellipses or perfect circles. Although the motions of comets, planets and their satellites could be *quantified* using Kepler's laws, what was the underlying force driving them, and could a theoretical proof of this force be arrived at? Realizing that the same unseen force exerted by the Sun was probably at work in both curving a comet's path and causing the planets to orbit the Sun, Newton set about defining precisely how this mysterious force worked. He called this force gravity (meaning 'weight').

Spurred on by the Great Comet of 1680, attempts to grasp the fundamentals in the post-Keplerian era were made in England by the likes of Robert Hooke (1635–1703), Christopher Wren (1632–1723) and Edmund Halley (1656–1742). In 1684 Halley famously asked Newton what would be the shape of the orbit of a body about the Sun if the force acting upon it was inversely proportional to the square of the distance. Without hesitation, Newton replied that the shape would be elliptical, but, unable to lay his hands on the calculations, he promised to look through his papers for proof. In November of that year Newton sent a manuscript to Halley, *On the motion of bodies in orbit*, containing those proofs; it was to form the nucleus of one of the greatest science books of all time, *Philosophiæ Naturalis Principia Mathematica* (*Mathematical Principles of Natural Philosophy*) often referred to as *The Principia*, which was eventually published in 1687 with the backing of Halley.

In *The Principia*, Newton introduced the universal law of gravitation. It holds that the force between any two bodies is proportional to the product of their masses and inversely proportional to the square of their distance from each other. The inverse square law of force explains the accelerating fall of an apple to the center of Earth, the acceleration of the Moon to Earth and how all the planets and comets orbited the Sun. Comets have so little mass that their gravitational effects on the planets is undetectable; however, the paths of comets themselves can be greatly affected by the gravity of other planets, especially when comets make close planetary flybys.

Interestingly, Newton correctly pointed out that observations of the swift, close passage past Earth of a comet whose orbit is well-known could be used to refine our estimations of the true size of the Solar System by means of parallax. Halley was the first to suggest that the same method of determining true distances in the Solar System could be applied to parallax measurements of a more predictable (but rare) event: the transit of Venus across the face of the Sun. Sadly, both Newton or Halley never saw such an event; the last was in 1639, before either was born, and the next pair of transits was due in 1761 and 1769, long after they had died.

The Principia also introduces Newton's laws of motion. Both Galileo and Descartes had surmised that a moving ball would travel forever at the same velocity in the absence of frictional forces. Newton's first law gives a clear definition of what a force does—a body remains at rest or moves with uniform motion in a straight line unless acted upon by a net external force. A net force acting on a body accelerates it either by changing its speed or its direction.

How forces can be measured is given by Newton's second law. Force is proportional to the rate of change of linear momentum. The acceleration 'a' of a body is parallel and directly proportional to the net force 'F' acting on it, is in the direction of the net force and is inversely proportional to the mass 'm' of the body, i.e., $F = ma$. From this definition we obtain the SI unit of force known as the newton (symbol 'N'): $N = kg\ m/s^2$—1 N will accelerate a mass of 1 kg by 1 m s^2.

Newton's third law states that if body 'A' acts on body 'B,' then body 'B' simultaneously acts on body 'A' with an opposite force of equal magnitude: $F\ AB = -F\ BA$. From Newton's second and third laws we can derive the law of conservation of linear momentum: the sum of the momenta before an interaction or collision is equal to the sum of momenta after a collision.

Newton's laws of motion can be applied to comets to calculate such things as the effects of forces other than gravity that might change a comet's path and its momentum; the effects of gaseous jets from the nucleus; collisions of the nucleus with other objects (including meteoroids, small asteroids and spacecraft); and the effect of the solar wind.

Halley's Insight

Newton's *Principia* specifically mentions the Great Comet of 1680 to illustrate the motions of objects in the Solar System. However, Newton considered that comets took parabolic paths—perfectly in accordance with his universal theory of gravitation—and he disagreed with Halley's suggestion that some of them may have elliptical orbits and, like planets, were permanent fixtures of the Solar System.

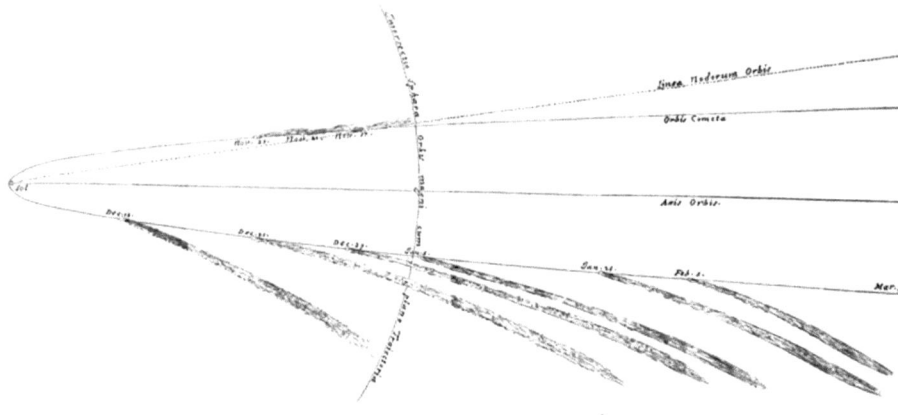

Fig. 3.7 This illustration of the inner part of the orbit of the comet of 1680–1681 featured as a large fold-out section in early editions of Newton's *Principia*

Nevertheless, Halley set about compiling a study of comets that had been observed through history—24 of them between 1337 and 1698—in an attempt to discover whether any of their orbits matched those of newly discovered ones. Halley's work took into account the fact that cometary orbits can be affected by the gravity of major planets, especially giant Jupiter, and consequently their orbital shapes and periods are subject to change over time. His painstaking research applied Newton's theory of gravity to some of the best observations available, including those by Cassini at Paris and Flamsteed at Greenwich, and was finally published in 1705 in his *Synopsis of Cometary Astronomy*. Halley wrote:

> By comparing together the accounts of the motions of these comets, 'tis apparent, their orbits are dispos'd in no manner of order; nor can they, as the planets are, be comprehended within a Zodiac, but move indifferently every way, as well retrograde as direct; from whence it is clear, they are not carry'd about or mov'd in vortices [suggested by Descartes]. Moreover, the distances in their perihelia are sometimes greater, sometimes less; which makes me suspect, there may be a far greater number of them, which moving in regions more remote from the Sun, become very obscure; and wanting tails, pass by us unseen.

Halley logically suggested that comets that had been seen in 1456, 1531, 1607 and 1682 (the latter having been observed by him) were returns of the same comet with a 76-year orbital period, and famously predicted that it would return in 1758. Halley died 16 years before the comet made its predicted reappearance. It was the first clear demonstration of the power of Newtonian gravity. The following year, Halley's pivotal role in finally admitting comets into the Sun's family was acknowledged by French astronomer Nicolas de Lacaille (1713–1762) when he named it Halley's Comet. Now designated 1/P Halley, it is the best-known of 278 known named short-period comets (those with orbital periods of less than 200 years and whose returns have been observed more than once) and the only one that becomes visible to the unaided eye on each return (Fig. 3.7).

Convinced that the Great Comet of 1680 was in an elliptical orbit around the Sun rather than an open-ended 'just passing through' parabolic path, Halley found that similar circumstances had attended three bright comets in the past—one in 44 B.C., another in 531 and the third in 1106—spaced at intervals of around 575 years, the same interval between 1106 and 1680–1681. After applying the same predictive method that he had used with the comet of 1682, he speculated that the Great Comet of 1680 would return around the year 2255. Although Halley was wrong on this score, incredibly the Great Comet of 1680 does share some of the orbital elements of comet C/2012 S1 (ISON), suggesting that the two comets, separated by 333 years, may be large fragments from the same giant parent comet.

In 1717 Halley inadvertently discovered his own comet while observing Mars. The 61-year-old astronomer noted:

> On the evening of Monday 10th June…the sky being very serene and calm, I was desirous to take a view of the disk of Mars (then very near the Earth, and appearing very glorious) to see if I could distinguish in my 24 foot telescope, the spots said to be seen on him. Directing my tube for the purpose, I accidentally fell upon a small whitish appearance near the planet, resembling in all respects such a nebula…The Reverend Mr. Miles Williams, Mr. Alban Thomas, and myself contemplated this appearance for above an hour…and we could not be deceiv'd as to its reality; but the slowness of its motion made us at that time conclude that it had none, and that it was rather a nebula than a comet…however, suspecting that it might have some motion, I attended the next night, June 11th, at the same hours and in the same company, when with some difficulty by reason of the thickness of the air, we found the two little stars, but the nebula could not at that time be seen, which we then imputed to the want of a clearer sky. But on Saturday, June 15th, the Moon being absent, and the air perfectly clear, we had again a distinct view of the two stars, with an entire evidence that there remained no footstep or sign of it, in the place where we had first seen this phenomenon, which we therefore now found to be a comet. And that being far without the orb of the Earth, and in itself a very small body, it appeared only like a little speck of cloud, such as would scarce have been discerned in an ordinary telescope, much less by the naked eye.

In the same letter for *Philosophical Transactions*, Halley speculated that "there may be still a much greater number of these bodies, which by reason of their smallness and distance are wholly invisible to the naked eye; so that unless chance do direct the telescope of a proper observer, almost to the very points where they are (against which there are immense odds) it will not be possible for them to be discovered."

Halley was right. Chance did (and still does) play a significant role in the discovery of new comets, but the chance of success is governed by a number of important factors. Although Halley had inadvertently discovered a comet while conducting a routine planetary observation—something that has been done many times since—most new comet discoveries made at the telescope eyepiece have been made by dedicated comet hunters working to a systematic plan of sweeping the skies through binoculars and telescopes. It is as true today as it was in the eighteenth century that the visual comet hunter is most effective when, by means of constant practice, he or she is thoroughly familiar with the appearance of star fields and the presence of any nebulae that may be mistaken for comets.

It's an inspirational exercise to review the efforts and achievements of some of the more prolific comet discoverers throughout history. This personal selection begins in the mid-eighteenth century, when comet-hunting really took off, and ends with the arrival of a technology that was to increasingly surpass visual hunters of comets and asteroids.

Charles Messier, the Ferret of Comets

One of visual astronomy's greatest comet hunters, French astronomer Charles Messier (1730–1817), had been inspired by his childhood views of one of the most spectacular comets of the eighteenth century—the Great Comet of 1744 (de Cheseaux's Comet, C/1743 X1), noted for its grand fan-like display of six tails. His drive towards astronomical discovery originated with his attempts to become the first astronomer to recover Halley's Comet at its predicted return in 1758. Even though that particular honor was not to be his, Messier found himself thoroughly hooked on comet hunting. Between 1758 and 1805 he observed no fewer than 44 comets, 13 of which were his own discoveries. Messier's impressive tally of observations earned him the nickname 'the ferret of comets' from none other than King Louis XV of France (Figs. 3.8 and 3.9).

The predicted return of Halley's Comet hadn't just induced an obsession for hunting comets in Messier alone. As we shall see, it ushered in an era of dedicated comet hunters across Europe (later spreading around the globe), most of whom were amateur astronomers eager to make a name for themselves on Earth as well as in the heavens.

Messier's name is, of course, forever linked with a list of fairly bright deep-sky objects—open and globular star clusters, nebulae and galaxies—that is still eagerly consulted by amateur astronomers today. The origin of the list dates back to August 28, 1758, when, during a sweep for a comet that he had been following for a fortnight (C/1758 K1, discovered by De la Nux in May of that year) Messier stumbled across a nebulous, comet-like patch in Taurus. After checking on its position some time later he found that the patch hadn't moved, and he realized that this was a newly discovered example of a dim nebula (Latin for 'little cloud'). Set permanently in starry realms, such nebulae (now considered deep-sky objects) had the potential to confuse comet hunters and lead to false announcements of cometary discoveries, so Messier decided to compile a catalog containing a complete list of all known nebulae, including any that he might subsequently discover.

Just 45 deep-sky objects were featured in the first version of Messier's catalog. Object number one—M1, the Crab Nebula—was the 'faint fuzzy' that had originally prompted him to compile his list. Exactly 103 objects were included in the final version, published in *Connaissance des Temps* (1781). During the twentieth

Fig. 3.8 Charles Messier, painted by Desportes in 1771

century eight more objects were added following research uncovering further observations of Messier and his friend Pierre Méchain (1744–1804). Méchain himself, also a prolific comet hunter, discovered 27 deep-sky objects, 18 of which made it into Messier's list. After gaining a few later additions, the list now totals 110 deep-sky objects. Although it is by no means an exhaustive catalog—it is restricted to the skies visible from Messier's location and it neglects to include a surprising number of reasonably bright and easily accessible deep-sky objects—it is still eagerly consulted today by amateur astronomers keen to observe many of the brightest deep-sky objects visible from the northern hemisphere.

Messier's criteria determining whether an object appeared nebulous were pretty broad. Some objects, such as the bright open star cluster the Pleiades (M45) in Taurus can only appear as a fuzzy nebula to the unaided eye if your eyesight isn't perfect; most people (especially young people) can see that it is made up of a number of stars. Although it's true that there is a little nebulosity to be glimpsed around parts of the Pleiades through large binoculars or in a rich-field telescopic view (given good seeing and dark skies), this was not Messier's reason for including it.

Fig. 3.9 Inspirational to young Messier, the Great Comet of 1744 displayed a wonderful *fan-shaped tail* array. From *The World of Comets* (1877) by Amédée Guillemin

Going a step further, the open cluster Praesepe (M44) in Cancer can't be resolved into individual stars with the unaided eye, but it readily breaks up into stars with virtually any optical aid.

There are other star clusters in Messier's list that didn't resolve into stars through his instrument—objects he genuinely thought to be nebulae—but which were later resolved through better instruments. In fact, Messier's list of 'nebulae' breaks down into an assortment of deep-sky objects. It features 1 supernova remnant (M1), 26 open star clusters, 29 globular clusters, 4 planetary nebulae, 7 star-forming nebulae, 40 galaxies (27 spiral, 4 lenticular, 8 elliptical and 1 irregular) and 3 others (the Milky Way patch M24, double star M40 and asterism M73).

Messier's record of cometary discovery is impressive. C/1760 B1 (Messier), the first of his 13 original comet discoveries, was visible for 2 months, and accurate positional measurements enabled it to become the 50th comet with a known orbit. Ironically, Messier just missed out on adding another deep-sky object to his list when, on February 12, 1760, the comet passed less than a degree east of the galaxy NGC 2903 (magnitude +9.6) in Leo, which went unnoticed.

On January 3, 1764, Messier's keen naked eye picked out his third comet, C1764 A1 (Messier) amid the stars of Draco when it was near the horizon and sporting a 4° tail. This wasn't to be his only naked-eye discovery; on April 8,1766, Messier sighted his fifth comet, D/1766 G1 (Helfenzrieder-Messier) in Aries, ready-made with an 8° long tail. Exactly a month earlier, Messier had discovered C/1766 E1 (Messier) by chance while following up curious reports that Venus

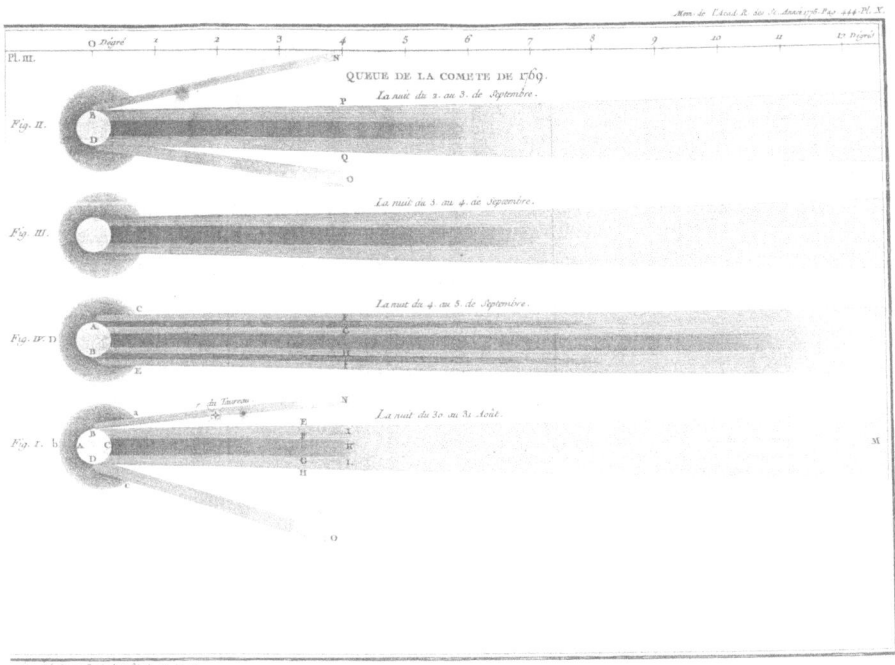

Fig. 3.10 Some of Messier's observations of the Great Comet of 1769, published in his *Grande Comète* (1808)

might have a moon of its own. Instead of a Venusian satellite he found a faint comet near the planet; another ironic twist came 2 days later when the comet passed within a degree of spiral galaxy M74 (magnitude +9.4) in Pisces, then undiscovered. Although M74 is in Messier's catalog, it had to wait another 14 years until Pierre Méchain would discover it.

C/1769 P1 (Messier), the sixth comet to be discovered by Messier, proved to be his greatest find. Discovered on August 8, the Great Comet of 1769 was initially reported as having 'appeared faintly in the telescope'; just a week later it had developed a 6° long tail and was visible with the unaided eye. On August 30 James Cook aboard the ship *Endeavour* (fresh from observing the Venus transit from Tahiti 2 months earlier) noted the comet in his logbook, stating that it had a tail 42° long. As its brightness increased, the comet's tail grew from a length of 60 to nearly 100° between September 9 and 11 as it made its closest approach to Earth. As perihelion approached, observations became more difficult, but the comet was recovered in late October and was followed until December (Fig. 3.10).

Six months after the Great Comet of 1769 had faded into the depths of space, Messier discovered one of the most remarkable comets in history—D/1770 L1 (Lexell). Messier's discovery was made on June 14, 1770, while viewing nebulae in Sagittarius, following a routine observation of nearby Jupiter. The fact that Messier spotted the comet, which was then very faint, among the rich star fields of the Milky Way, demonstrates his great familiarity with the sky and his eye for anything appearing to be out of place—a trait shared by all of history's great visual comet discoverers.

Ten days later D/1770 L1 had brightened to magnitude +2 and its coma had ballooned to the apparent size of the Moon. On July 1 the coma had grown to a giant 2½ degrees across—five times the Moon's apparent diameter—and was moving exceedingly rapidly across the sky. With a bright nucleus reported to be the apparent size of Jupiter, English astronomer James Six reported that the comet had traced a path 42° long in the previous 24 h—so fast that its motion against the stars at this time could have been be perceived in real time through the telescope. Such circumstances only attend objects that make a very close encounter with Earth, and sure enough, orbital calculations showed just that—on July 1 it passed just 0.015 AU from Earth. That's a mere 2.2 million km, or less than six times the Moon's average distance. As such, D/1770 L1 holds the record for the closest observed cometary approach to our planet.

Following perihelion, observations of the fading comet continued throughout August and September. Messier himself was the last person to observe D/1770 L1, when on October 3 he reported it as being "extremely faint and very difficult to observe." The comet was never seen again, and it is considered lost.

Later advanced orbital calculations made by Swedish astronomer Anders Lexell (1740–1784) determined that when observed the comet had an orbital period of 5.58 years, making it the comet with the shortest period then known. But D/1770 L1 was different. It had never been seen previously because Lexell showed that 3½ years before its discovery its orbit had been dramatically shunted into a smaller ellipse while making a close pass by Jupiter in January 1767. Working in conjunction with Pierre-Simon Laplace (1749–1827), Lexell showed that the comet had again encountered Jupiter in 1779, perturbing its orbit once more and flinging it into the remote depths of space, perhaps never to return. Lexell's work on determining the remarkable journey of D/1770 L1 was enough to earn him recognition in the comet's name, rather than its discoverer, Messier.

D/1770 L1 was the first known observed Jupiter family comet—one with an orbital period of less than 20 years and an inclination of less than 40°—although it was very short-lived in that category. In addition, it was the first identified near-Earth object (NEO). For around a hundred years astronomers had known that comets were celestial objects in orbit around the Sun, and that close passes to Earth were inevitable, given enough time. The close approach of D/1770 L1 gave astronomers pause to think. What if Earth passed through a comet's tail, or what if a comet were to actually collide with our planet? (Fig. 3.11).

C/1785 A1 (Messier) is the final comet with which Messier is credited as the sole discover. His last co-discovery—C/1801 N1 (Pons-Messier-Méchain-Bouvard)—took place on July 12, 1801, when he was aged 71. As its name suggests, this comet is remarkable for the fact that it was discovered by four independent astronomers at almost exactly the same time as each other.

A summary of our understanding of the nature of comets in Messier's time was given by philosopher Charles Bonnet (1720–1793). "But what are comets?" he

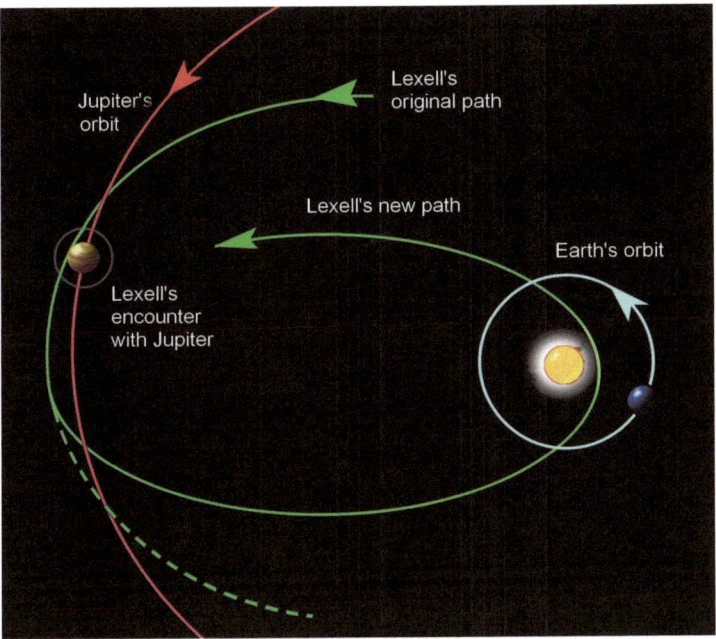

Fig. 3.11 In 1767 Jupiter's gravity perturbed Lexell's Comet from its original orbit into a smaller Earth-crossing ellipse (Illustration by the author)

asked. "Planets not fully formed; or planets destroyed by conflagration? Or bodies of an wholly different nature, of which, therefore, we can form no idea?"

Bonnet described them as:

> ...another class of planets appertaining to our system, and which exhibit appearances vastly different from any of the other planets. The nucleus or star, seems much dimmer; they are to appearance surrounded with atmospheres of a prodigious size...What is called the head of a comet, or the round part, appears to be a solid globe, and is called the nucleus, which is easily distinguished from the atmosphere of hairy appearance...although the orbits of all comets are very eccentric ellipses, yet there are very great differences among them. Excepting the orbit of Mercury, there are no great differences among those of the planets, either as to their eccentricity, or the inclination of their planes; but the planes of some comets are almost perpendicular to others, and some of their ellipses are much wider than others...It is now universally allowed that comets are opaque bodies, enlightened by the Sun...The extraordinary atmospheres, or tails of comets, have given rise to various conjectures, though it is allowed by all, that they depend on the Sun some way or another.

Caroline Herschel, the Comet-Sweeper

Caroline Herschel (1750–1848), sister of the great observational astronomer William Herschel (1738–1822), was one of the few women active in astronomy during the period in which she lived, a situation brought about by the prevailing

Fig. 3.12 Caroline Herschel, depicted in 1829 by Tieleman

societal norms of the era. Although she assisted her brother during his remarkable observational career, she was encouraged by him to take a more independent role in astronomy. In 1783 he presented her with a 27-in. focal length Newtonian telescope with which she could pursue her own studies. She was to become in her own right an extremely capable amateur astronomer who made a number of significant discoveries at the telescope eyepiece, including a number of comets (Fig. 3.12).

Between 1786 and 1797 Caroline Herschel discovered eight comets, six of which she had unquestioned priority, namely: C/1786 P1 (Herschel); 35P/Herschel-Rigollet (1788); C/1790 A1 (Herschel); C/1790 H1 (Herschel); C/1791 X1 (Herschel); C/1797 P1 (Bouvard-Herschel). Both C/1793 S2 (Messier) and 2P/Encke (1795) were found independently (but not originally) by Herschel. Several of these are worth a look at in more detail (Fig. 3.13).

35P/Herschel-Rigollet was discovered by Herschel on December 21, 1788, 1° south of Beta Lyrae and visible only through a telescope. The comet was found to have reached perihelion around 4 weeks before, and it was followed until February 5, 1789. We jump forward more than a century and a half to July 28, 1939, when Roger Rigollet discovered a telescopic comet that was soon found to have an identical orbit with Herschel's comet of 1788. Last observed on January 16, 1940, from Lick Observatory, 35P/Herschel-Rigollet won't come back around until the late twenty-first century.

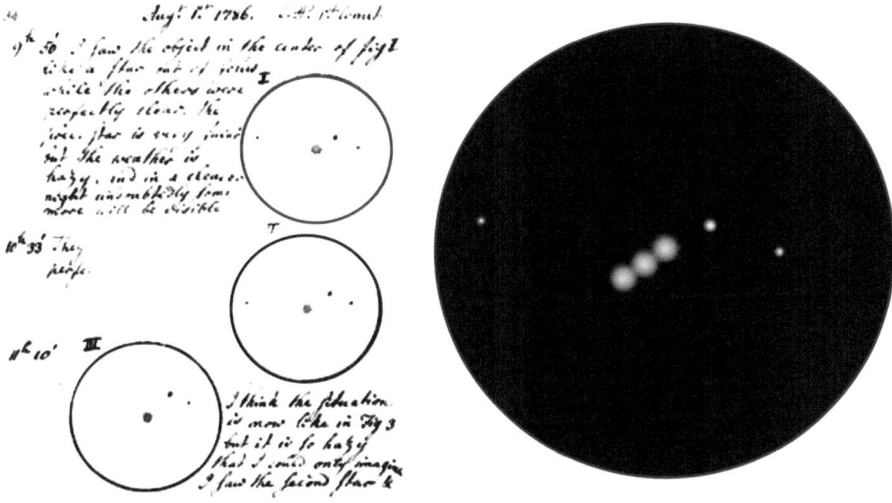

Fig. 3.13 Caroline Herschel's first cometary discovery, C/1786 P1, as depicted in her notebooks and showing movement through the telescopic field of view between 9.50 and 11:10 p.m. on 1 August 1789, along with a simulation showing the comet's combined motion (Illustration © by the author)

C/1790 H1 (Herschel), discovered on April 18, 1790, as a tail-less 7th magnitude patch, was to become Herschel's second brightest comet; by perihelion on May 21 it had developed a 4° long tail and shone at the 4th magnitude. More than 20 months were to pass before anyone discovered another comet. The dearth of comets was broken by Herschel herself when, on December 15, 1791, she found C/1791 X1 (Herschel). Shining at the 6th magnitude, the comet, which she described as being 'a pretty large telescopic comet,' reached perihelion on January 14, 1792, at a distance of 1.29 au, but increasing in distance from Earth the comet faded and was last seen on January 28 by Charles Messier.

Herschel's final comet discovery, C/1797 P1 (Bouvard-Herschel), proved to be the brightest of her tally. She found the comet from Slough, England, on August 14, 1797, within hours of Alexis Bouvard (1767–1843) in Paris. Already a 3rd magnitude object, the comet passed just 13 million kilometers from Earth 2 days later, making it the 15th closest known cometary approach in history. An easy naked-eye object, it appeared as a round nebulosity with no distinct nucleus. C/1797 P1 faded quickly and was last seen by Bode on August 31.

Caroline Herschel's tally of cometary discoveries was to stand as a record for women astronomers for 186 years, when on May 27, 1984, the American astronomer Carolyn Shoemaker found her sixth comet, C/1984 K1 (Shoemaker). Shoemaker (born 1929) went on to surpass Herschel and holds the record for cometary discoveries—no fewer than 32 of them, discovered on photographs taken with the 40 cm (16-in.) wide-field Schmidt telescope at the Palomar Observatory.

Fig. 3.14 Contemporary illustration of Jean-Louis Pons painted by Ernesto Buonoguli, 1829

Pons, Le Grand Chercheur

After making a humble start in 1789 as caretaker at the Observatory of Marseilles, Jean-Louis Pons (1761–1831) was to follow in the footsteps of his compatriots Messier and Méchain by becoming a prolific discoverer of comets during the first quarter of the nineteenth century. His tally of 37 comets discovered at the telescope eyepiece between 1801 and 1827 earns him the accolade of being history's greatest visual comet discoverer (Fig. 3.14).

Although Pons came from a poor family and had received very little formal education, and was then at the relatively advanced age of 27, the academic staff at Marseilles realized that his passion for astronomy was infinitely greater than his commitment to the unskilled duties for which he had been employed. As a non-scientific assistant he lapped up valuable lessons from professional astronomers at the observatory and, over time, acquired skills that made him a respected astronomer in his own right. While assisting astronomers with their observations he particularly impressed them with the ability to mentally picture star fields and to notice changes in their appearance. Perhaps unsurprisingly, this is a skill common to many great comet discoverers throughout history.

Pons swept the skies using his favorite instrument, Le Grand Chercheur, a self-designed short focal length telescope with a generous 3°-wide actual field of view, and made his first comet discovery on July 11, 1801. Sharing the find with three other astronomers, C/1801 N1 (Pons-Messier-Méchain-Bouvard) proved to be the last that bore Messier's name as a co-discoverer. Pons' observational work was rewarded in 1813 when he was promoted to Assistant Astronomer at Marseilles, becoming the observatory's assistant director.

Despite his growing status, Pons' lingering rustic naïveté prompted astronomer von Zach (founder of the 'Celestial Police', see above) to attempt a leg-pulling—a joke that, famously, proved immensely beneficial to Pons. In response to a letter from Pons bemoaning the dearth of recent cometary discoveries, von Zach advised him to redouble his comet-hunting efforts when the sunspot count increased. Of course, von Zach considered that his 'advice' was harmless fun at Pons' expense, since he thought that there was no connection between solar activity and the frequency of comets. But Pons treated von Zach's words seriously, and a while later he wrote him a grateful letter. He'd found his first comet in many months, shortly after the Sun had developed a group of large spots, just as von Zach had predicted. Indeed, Pons attributed his success as a prolific discoverer of comets to this very advice.

We now know that there is indeed a connection between solar activity and cometary phenomena. Solar radiation pressure acts on the nucleus and helps form their tails; 'gusts' in the solar wind can produce enhanced activity in a comet's nucleus, the interaction between magnetic fields in the solar wind, and a comet's plasma tail can lead to dramatic changes in its appearance, even producing tail-decoupling events.

Pons played an important role in the discovery of comet 2P/Encke, the second periodic comet to have been discovered. The comet had been observed by Méchain in January 1786, by Caroline Herschel in November 1795 and by Pons in October 1805, but on each occasion it had been mistaken for a new comet. Heinrich Olbers suggested that they were one and the same object. Pons 'discovered' it again in November 1818; reaching naked eye visibility, it was observed for nearly 2 months, enabling a series of accurate positional measurements to be taken. After analyzing the observations, Johann Encke found that they corresponded with a single comet in an elliptical orbit with a period of 3.3 years—by far the shortest cometary period then known—and he predicted that it would next reach perihelion in May 1822. The comet was duly recovered in June 1822 by Karl Rümker from New South Wales, Australia.

After making a hat trick of cometary discoveries in 1818, the French Academy of Sciences awarded Pons the prestigious Lalande Prize. Eager to bask in the glory of Pons' cometary successes, Maria Luisa (Duchess of the Italian state of Lucca) appointed Pons to the directorship of a new observatory at the Royal Park La

Marlia, near Lucca in Tuscany. His title was Her Majesty's Astronomer Royal, Director of the Astrotropic Department of the Observatory, and Emeritus Professor of the Royal Lyceum. It didn't really take the royal cash prize that was offered for new discoveries from the observatory for Pons to set to work with his eagle-eye. Near the end of 1823 Pons discovered his brightest comet, C/1823 Y1 (De Bréauté-Pons); a naked-eye object when discovered, he initially mistook it for chimney smoke wafting above the horizon. Known as the Great Comet of 1823, it was visible throughout the first few months of 1824, striking its viewers with two tails, one pointing away from the Sun and the other pointing towards it. Pons was the last to view the comet, on April 1, 1824.

Despite its promising start, funding for the observatory at La Marlia dried up and the observatory closed. Royalty again favored Pons when, in 1825, he was offered the directorship of the Observatory of the Museum for Physics and Natural History by Leopold II (Grand Duke of Tuscany), where he found his final seven comets.

Although Pons was one of history's most energetic comet discoverers, and lived in an age when his exceptional visual acuity (until his eyesight deteriorated in the last couple of years of his life) was prized, he appears to have been somewhat lax in making clear, detailed notes and observational drawings. Nor was he much interested in following up his discoveries by making positional measurements—he left that to others. In modern terms he would be considered a less than diligent observer. Yet his tally of 37 visual comet discoveries is unsurpassed and is likely to remain so.

Edward E. Barnard, 'Eagle-Eyes' Surpassed

It's ironic that Edward Emerson Barnard (1857–1923), an astronomer noted for his particularly keen visual acuity, was to make the first photographic cometary discovery. His initials earned him the nickname of 'Eagle-eyed Barnard'—and rightly so, for in the decade between 1881 and 1891 he discovered 13 comets (3 of which turned out to be periodic) and co-discovered 2 more (Fig. 3.15).

On October 13, 1892, Barnard used the 6-in. Willard lens at the Lick Observatory to expose a photographic plate to image the Milky Way west of Altair. On examination the following day, the 4 h and 20 min exposure clearly showed a fuzzy trail near the center, about a quarter of a degree long. That night Barnard visually confirmed the comet as a small magnitude 13 patch at the eyepiece of the 12-in. Clark refractor. Slowly fading, it was last observed on December 8. Despite calculations pointing to a 6.52 year orbit, D/1892 T1 (Barnard 3) as it was known, failed to turn up in a number of searches during the twentieth century.

On October 7, 2008, Andrea Boattini accidentally discovered a faint 18th magnitude comet on a CCD image taken with the 28-in. (70 cm) reflector at Mount Lemmon Observatory, Arizona. After securing additional images, the comet was found to have an elliptical orbit with a period of 5.92 years. Further calculations showed that it was a perfect match for Barnard 3. In the period during which the comet was lost it had completed 20 orbits and had made close approaches to Jupiter in 1922, 1934 and 2005. The comet is now designated 206P/Barnard-Boattini.

Fig. 3.15 E. E. Barnard in 1895

Unlike the impact that photography made on asteroid discovery (discussed below), photography didn't render the visual comet hunter entirely redundant. Even today there is ample opportunity for the dedicated observational astronomer (or, in some cases, the fortunate casual sky-watcher) to make a comet discovery of their own (Fig. 3.16).

Towards a Modern View

In terms of the computation of cometary orbits (and of orbits in general), we've seen that great advances were made in the mathematical work of Newton, Halley, Lexell and Gauss (the latter's work is described below). It's fair to say that by the twentieth century astronomers were able to compute cometary orbits pretty well by modern standards, providing the astrometric measurements of cometary positions were accurate. Such measurements, once performed using micrometers and the

Fig. 3.16 As well as having exceptional visual acuity, Barnard was a skilled and accomplished early astrophotographer. This is his image of Comet Morehouse of 1908

observer's eye at the telescope eyepiece, were by the early twentieth century made using photographic plate measuring devices.

While cometary orbits had been pretty much pinned down by the nineteenth century, the nature and origin of the comets themselves remained a matter of wide-ranging speculation. In 1948 the British astronomer Raymond Lyttleton (1911–1995) suggested that comets were formed at the dawn of the Solar System as loosely packed swarms of ice and dust, interstellar dust and gas. Just 2 years later, Lyttleton's 'flying sandbank' model was to engage in competition with the 'dirty snowball' theory of American astronomer Fred Whipple (1906–2004).

Interestingly, as an aside, the comet debate was contemporaneous with the transatlantic debate over the origin of the Moon's craters. The British, exemplified by the amateur astronomer Patrick Moore (1923–2012), generally favored the volcanic theory, vs. the United States, represented by Ralph Baldwin (1912–2010), a scientist whose impact theories were based on close observation and experimentation. Again in this case, the United States won the debate in the late twentieth century when Apollo directly sampled the Moon's surface and brought back indisputable evidence that the majority of the Moon's craters were formed through the impact of asteroids and comets.

The fact that cometary paths often refused to keep to a strict Newtonian schedule suggested to Whipple that the nucleus was solid chunk of water ice and other volatiles in which dust grains were embedded; when sublimating under the Sun's heat, jets of gas would produce a reaction effect, slightly altering the comet's velocity and path. The flowing gases would carry with them dust grains into space, which, traveling faster than the nucleus' escape velocity, would produce a dust tail. Whipple's model was vindicated in spectacular fashion on March 13, 1986, when the ESA probe *Giotto* encountered Halley's Comet, passing the nucleus by just 596 km and returning spectacular images of a dark, solid, 'icy mudball.'

Chapter 4

Comets of the Enlightenment: Great Comets of the Seventeenth and Eighteenth Centuries

Only under exceptional circumstances can any of the periodic comets appear bright enough to be easily seen with the unaided eye, and they rarely assume spectacular proportions in the night skies. A relatively small, intrinsically faint comet can appear bright and impressive in the night sky if its orbit brings it into close proximity with Earth; conversely, a far bigger, intrinsically brighter comet can appear small and faint if its perihelion takes place on the opposite side of the Sun to Earth. Optimum circumstances for any comet to occur is when it is at its intrinsically brightest near perihelion and in close proximity to Earth at the same time; a sizable apparent angular distance from the Sun at this time will mean that it will be seen in all its glory against a fairly dark sky background.

Most newly discovered comets never develop into anything other than fairly dim objects that can only be faintly glimpsed through binoculars or telescopes. Now and again, say once or twice in a decade, a comet bright enough to be viewed with the unaided eye becomes visible. Really spectacular bright comets—known as Great Comets—are rare, with only a handful coming into view every century.

C/1618 W1

First sighted in late November 1618, several weeks after perihelion (0.4 au from the Sun) on November 8, this was the first Great Comet to have been observed since the invention of the telescope—indeed, it was the first comet to have been viewed through a telescope. C/1618 W1 was a prominent sight from the northern hemisphere as it passed through Libra, Virgo, Boötes (where, on December 6, it was at its closest to Earth at 0.36 au), Ursa Major and Draco during December.

P. Grego, *Blazing a Ghostly Trail: ISON and Great Comets of the Past and Future*,
The Patrick Moore Practical Astronomy Series, DOI 10.1007/978-3-319-01775-4_4,
© Springer International Publishing Switzerland 2014

Fig. 4.1 Contemporary depiction of the Great Comet of 1618, from a Frankfurt (German) publication of 1619

Having first sighted it through a break in the morning clouds over Linz (Austria), Johannes Kepler followed the comet as it headed north, taking measurements of its position up to January 7, 1619. The true brightness of C/1618 W1 is difficult to gauge owing to the antiquity and variety of observational sources, but it is likely to have been around magnitude 0 to 1 at its brightest. Far eastern sources report a bright, tailed comet that gave off a 'blue-white vapor' as it passed through Ursa Major. Observing from Ingolstat (Germany), Johannes Cysat sketched the length of the tail as 55° long (December 8) and 70° long (December 9) (Fig. 4.1).

C/1664 W1

First sighted in mid-November 1664, a week after perihelion (December 4, 1.03 au) this comet was to become a glorious sight towards the end of 1664, passing within 26 million km of Earth (December 29, 0.17 au) and attaining a brightness of magnitude −1. During December the comet's path took it from Corvus, through Hydra, Antlia, Pyxis, Puppis, Canis Major, Lepus and Eridanus (fairly low to the horizon for the UK and better seen from southern Europe and America), and then northward into Taurus, Cerus and Aries during the first half of January. It was last seen on March 20, 1665.

The first Great Comet of the Enlightenment, C/1664 W1 was viewed by all the great scientists and thinkers of the Western world, including Newton, Halley, Hooke, Pepys, King Charles II, Hevelius and Cassini (to name but a few). Indeed it is thought to be the comet that is referenced by John Milton in *Paradise Lost, Book II* (1667).

On the other side of the Atlantic, in the burgeoning colony of New England, Samuel Danforth was so impressed with the comet that he wrote a book about it: *An Astronomical Description of the Late Comet or Blazing Star; As it appeared in New-England in the 9th, 10th, 11th, and in the beginning of the 12th Moneth, 1664. Together with a Brief Theological Application thereof* (1665) was one of the first astronomical works ever published in America. Many of Danforth's postulations about the nature of the comet were correct, showing that he was familiar with the latest knowledge and speculations of European astronomers. Danforth postulated that the comet was a heavenly body more distant than the Moon, that its coma and tail (which pointed away from the Sun) were reflections of sunlight off material streaming from the comet's head, and that it had a smooth motion along a uniform curve. As well as being an astronomer and almanac-maker, Danforth was a Puritan minister, and he lost no opportunity to tie this celestial apparition with the will of God, taking it to be a divine warning to New Englanders to keep to the path of moral righteousness (Fig. 4.2).

C/1665 F1

Following hard on the heels of C/1664 W1, late March 1665 saw the appearance of another Great Comet. Swinging by Earth on April 4 (0.57 au), C/1665 F1 was visible to the naked eye through April 20, when it had attained a brightness of around magnitude −1. This was several days prior to perihelion on April 24 (0.11 au). Famously observed by Hevelius, an analysis of the comet's motion featured in his *Cometographia* (1668) as it moved from Cygnus through Pegasus to Andromeda.

In *De Cometis* (1665) John Gadbury was particularly keen to broadcast the supposed ill omens that comets bring. "Threatening the world with famine, plague and war. To Princes, death! To Kingdoms, many crosses. To all estates, inevitable losses. To Herdsmen, rot. To ploughmen, hapless seasons. To sailors, storms. To Cities, civil treasons." Not long afterwards the plague came to England, but it isn't known whether Gadbury was pleased to see his predictions come true. It was while sheltering from the plague at Woolsthorpe Manor, his country home, that Isaac Newton was to experience a series of the most astonishing scientific insights in his tory (Figs. 4.3 and 4.4).

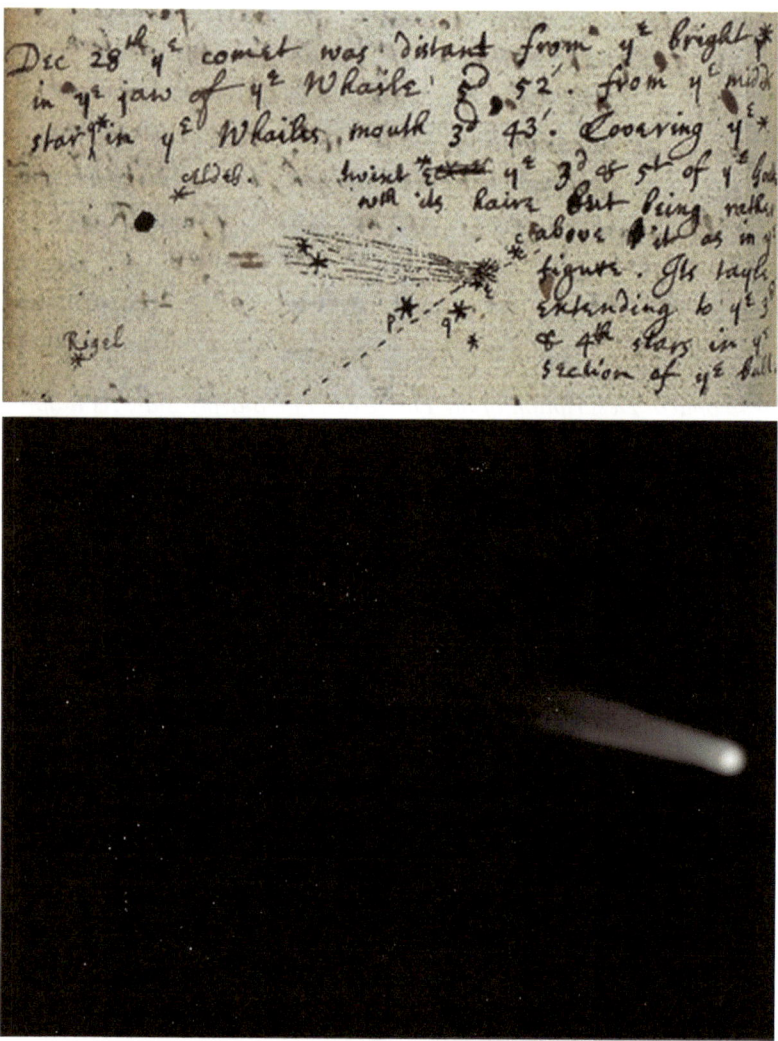

Fig. 4.2 Newton's sketch of C/1664 W1 (his first comet) made in his notebook on December 28, 1664, and (below) a simulation of what he saw, drawn by the author

C/1668 E1

Having come to perihelion around February 29, 1668, when it had passed within some 1.5° of the Sun, the Great Comet of 1668 was first 'officially' detected from the Cape of Good Hope, South Africa, on the evening of March 3 when in Cetus. A description of the comet, as seen on March 5 from Lisbon, Portugal, by an

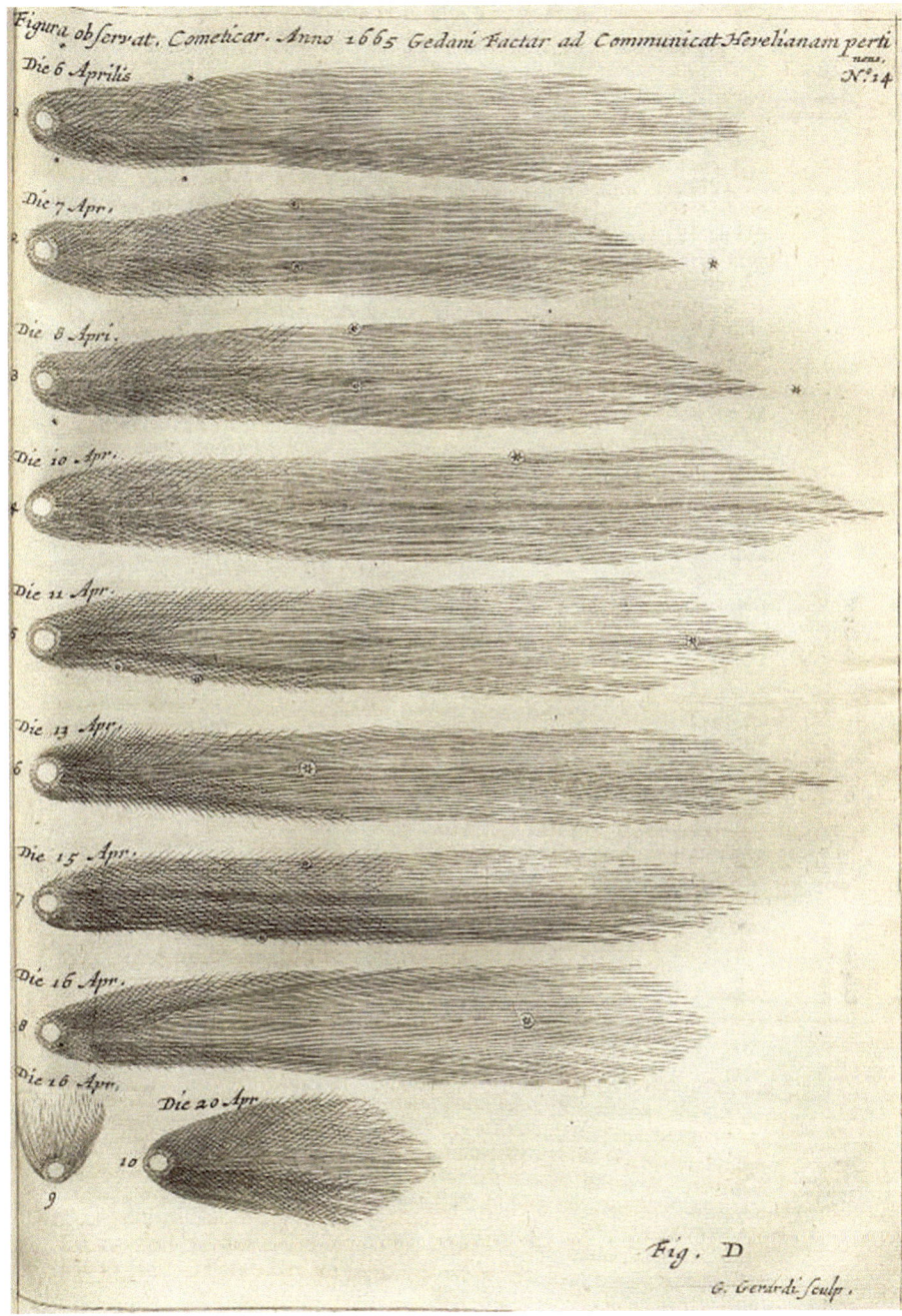

Fig. 4.3 Sequence of sketches by Hevelius showing the development of C/1665 F1, from *Cometographia* (1668)

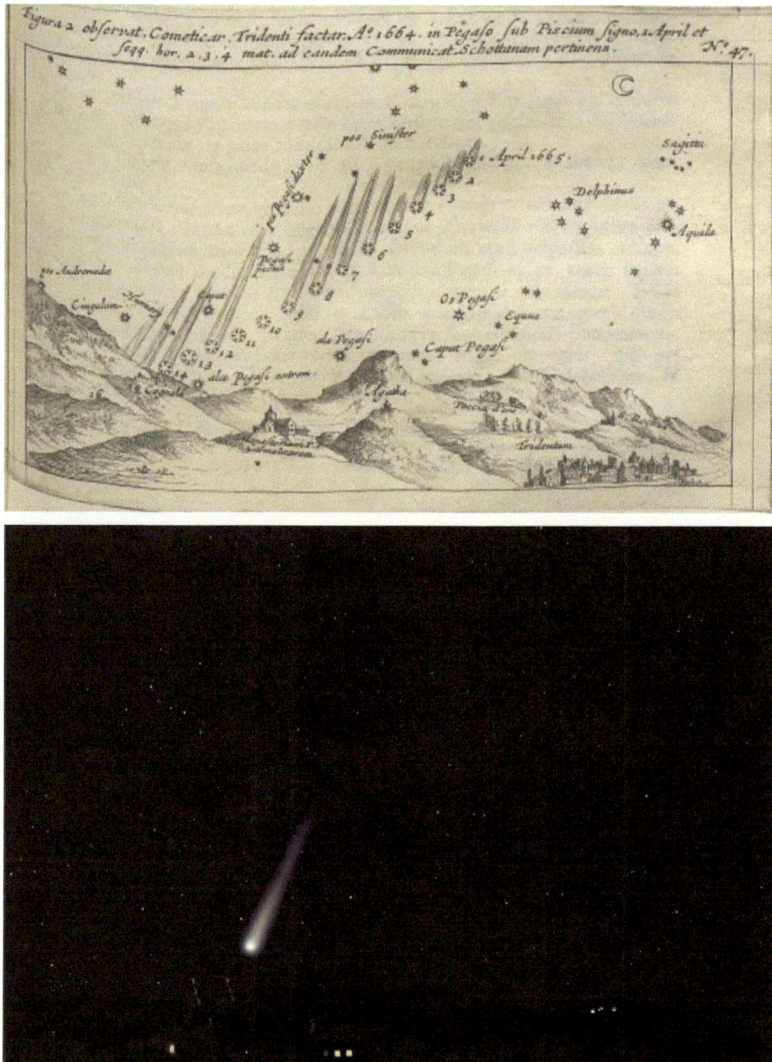

Fig. 4.4 Sequence of sketches by Hevelius showing the development of C/1665 F1, from *Cometographia* (1668) along with a simulated view of the comet as seen on the morning of April 12, 1665 at 4 a.m. Simulation drawn by the author.'

unknown observer, tells us that "the body [head of the comet] thereof is not seen, because it remains hid in the horizon. Its train is of a stupendous length, extending in appearance over almost a quarter of the visible heaven, from west to east; its apparent breadth is of a good palm [around 7°], and its splendor very great, but it lasts but a few hours."

Observing from much further south in San Salvador, Brazil, Valentin Estanzel sighted the comet on the evening of March 5, 1668. According to Estanzel, local people had reported sighting a comet in the morning skies the previous month, and it was assumed to have been the same object that was now so prominent at dusk. He reported that "[the comet appeared] a little above the horizon from west to east-southeast and it exhibited extraordinary brightness." Displaying a tail 23° in length, he remarked that "the 'globe' or head of it was so small and thin, that very few could discern it with the naked eye…As to the color of this comet, it was at first very splendid, and cast itself with that vividness upon the sea, that the rays thereof were reverberated unto the shore where the observers stood. But the brightness lasted only for 3 days, after which it did considerably decay. But that which seemed somewhat strange was, that having lost so much of its light, yet its bulk was not diminished, but continued rather increasing until the comet disappeared." After reaching its most southerly declination on March 6, Estanzel said that it had faded and its tail had become so translucent that "the eye could easily see the stars that were behind it." Chinese observers saw it as a 'stretch of white light' in the south-west, with a southeast-pointing tail more than 6° long.

From Bologna, Italy, Giovanni Cassini first observed C/1668 E1 on the early morning of March 10, noting that while the comet's head was obscured near the horizon its tail extended as a 'path of light' from Cetus into the neighboring constellation of Eridanus, the tail terminating near 14 Eridani. On the following evening he commented that he saw "a brightness in the Whale [Cetus], at least for half an hour, which was very like the splendor of Venus, likewise veiled by thin clouds," and by the 12th he remarked that the comet's tail extended away from the Sun and spanned some 32°. Although in the following days the comet faded somewhat, its tail was estimated by Chinese observers at 40° long when in Eridanus on March 18, confirming Estanzel's comments. Chinese accounts relate the last observations of C/1668 E1 on March 30 (Fig. 4.5).

C/1680 V1

After its discovery by Gottfried Kirch from Koburg, Germany, on November 14, 1680—the first comet ever to have been found at the eyepiece—C/1680 V1 went on to become the Great Comet of 1680, one of the brightest comets of the seventeenth century. It was this comet that prompted Isaac Newton to test his burgeoning ideas on gravity, and it was extensively followed for 4 months by astronomers in Europe, including Giovanni Cassini at the Paris Observatory and John Flamsteed at the Royal Observatory, Greenwich, as well as by astronomers in dominions further afield.

C/1680 V1 passed 0.42 au (68 million km) from Earth on November 30, and by December 4, it had brightened to magnitude +2 and sported a tail 15° long. By December 16, 2 days before perihelion, Matheus Merian wrote that when the head of the comet was on the horizon the tail "stretched from southwest to northeast and covered the most part of the sky. The tail was bright and clear, but transparent, so that the fixed stars underneath it were clearly visible and recognizable."

Fig. 4.5 Simulated view of the Great Comet of 1668 as observed by Cassini, looking west over Bologna at dusk on March 12 (Illustration © by the author)

A Sungrazer, the comet passed just 0.01 au (1.5 million km) from the Sun at perihelion, when it was reported to have been visible during the daytime; rounding the Sun in what was a hairpin turn, the comet emerged from the solar glare as a brilliant evening object sporting a very long tail. In late December, it reappeared in the evening sky in the west, again shining at magnitude +2, and had developed a long, slender golden tail that stretched for at least 70°, making it around 150 million km long.

A description of the comet's apparition, as observed from Mexico, was provided by Ivan Ratkaj:

[The comet] showed up for the first time at the end of November at four o'clock in the morning in Virgo with a long tail extending towards the west. The tail was dim at first, but brightened from day to day. The motion of the comet was directed from west towards the east, but so fast that after two days it was forty degrees to the east and closer to the Sun. After three or four weeks, when the comet passed the southern hemisphere, it showed up again after sunset with a horrifying tail which had a length up to fifty degrees; the body of the advancing comet was small and now moved towards the west and the tail was oriented towards the east for some time, and quickly afterwards towards the north...everything faded slowly away.

According to John Fiske (*The Dutch and Quaker Colonies in America*, 1903):

Late in the autumn of 1680 the good people of Manhattan were overcome with terror at a sight in the heavens such as has seldom greeted human eyes. An enormous comet, perhaps the most magnificent one on record, suddenly made its appearance. At first it was tailless and dim, like a nebulous cloud, but at the end of a week the tail began to show itself and in a second week had attained a length of 30 degrees; in the third week it extended to 70 degrees, while the whole mass was growing brighter. After five weeks it seemed to be absorbed into the intense glare of the Sun, but in four days more it reappeared like a blazing Sun itself in the throes of some giant convulsion and threw out a tail in the opposite direction as far as the whole distance between the Sun and the Earth.

Fig. 4.6 Contemporary painting by Lieve Verschuier showing the Great Comet of 1680 over Rotterdam

Having faded from naked-eye visibility by early February 1681, the comet was kept under observation well into 1681 and was last seen on March 19. Interestingly, the orbital elements of Comet ISON bear such similarity with the Great Comet of 1680 that there is reasonable speculation that the two may once have been a single comet that fragmented sometime in the distant past (Figs. 4.6 and 4.7).

1P/Halley 1682

On a number of occasions through recorded history, Halley's Comet has made a truly spectacular showing in the skies, placing itself in the league of Great Comets. This has happened twice in the period we're examining, giving a magnificent display during its apparitions of 1682 and 1910.

In 1682 the comet was first spotted by Georg Dörffel on August 15, and it remained a naked eye object for 41 days, attaining a maximum brightness of around magnitude 0 on August 31 when at its closest to Earth at a distance of 0.42 au (68 million km). Perihelion took place on September 15, some 0.58 au (87 million km) from the Sun.

Among viewers of the Great Comet of 1682 was 25-year-old Edmund Halley who, 13 years later, was to discover that the comet's orbit was remarkably similar

Fig. 4.7 Christoph Weigel's impression of the Great Comet of 1680

to comets that were seen in 1531 and 1607—so similar, in fact, that Halley speculated in his "Synopsis on Cometary Astronomy" (1705) that these were one and the same object and predicted its return in 1758.

It was at dawn on August 14, from near Hunting Creek, Maryland, looking west over the Patuxent River, Maryland, that Arthur Storer (an acquaintance of Isaac Newton) first set eyes upon the Great Comet of 1682. One of the first astronomers in the colonies, Storer (originally from Lincolnshire, England) kept a meticulous record of the comet until September 18, providing such an accurate account that it was to be referred to in Newton's *Principia* (Fig. 4.8).

C/1686 R1

Approaching the inner Solar System from the south, C/1686 R1—the sixth and final Great Comet to appear in the late seventeenth century—was discovered from Cape Town, South Africa, on August 12, 1686, by Simon van der Stel, who reported: "This night appeared in the fifth house of the heavens, at 1 o'clock, in the horizon a comet corresponding in length with Saturn and Venus conjoined, on the left shoulder of the Hare [near Beta Lupi]...the tail extended right, east and west to the length of 35 celestial degrees, in Gemini."

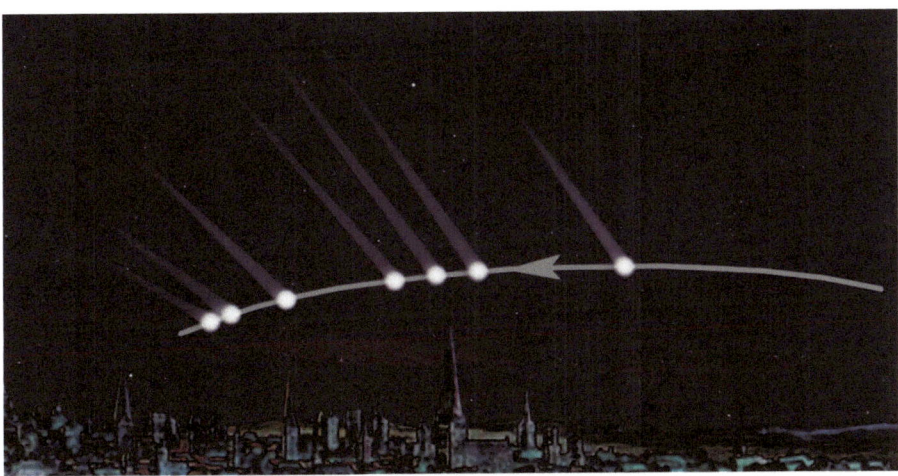

Fig. 4.8 View over London, looking west at around 8 p.m., showing a simulation of Edmund Halley's observations of the Great Comet of 1682 on August 26, 29, 30 and 31, and September 4, 8 and 9. The comet was later to bear Halley's name after he predicted its return in 1758. Simulation drawn by the author

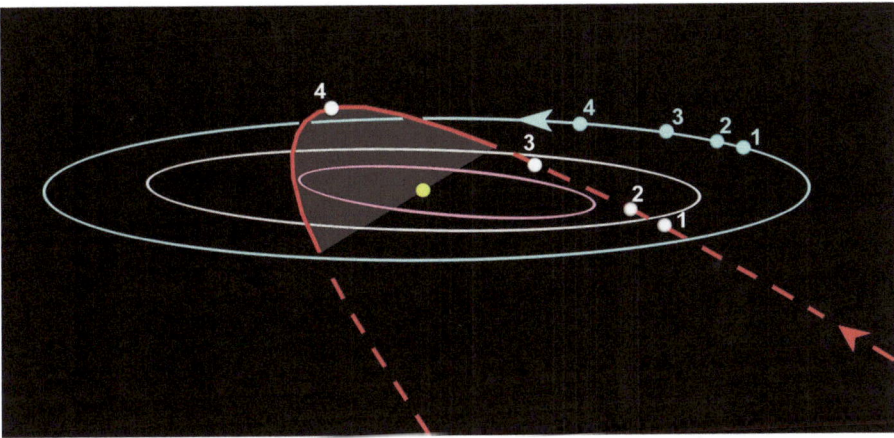

Fig. 4.9 The Great Comet of 1686 showing its path through the inner Solar System (*red line*: above ecliptic; *dashed red line*: below ecliptic) along with the orbit and positions of Earth (*blue*) and the orbits of Venus (*gray*) and Mercury (*pink*). Key: (1) August 12; (2) August 16; (3) August 27; (4) September 16 (Illustration by the author)

Heading north, the comet was visible with the unaided eye for 34 days. Perigee took place on August 16 at 0.32 au (48 million km), and it attained its maximum brightness on August 27, shining at the first magnitude. Perihelion occurred on September 16 at 0.34 au (51 million km) from the Sun (Fig. 4.9).

C/1743 X1

First sighted by Jan de Munck from Middelburg, Netherlands, on November 29, C/1743 X1 was also independently picked up by Dirk Klinkenburg at Haarlem, Netherlands, on December 9 and by Jean-Philippe de Chéseaux from Lausanne, Switzerland, on December 13. Sometimes referred to as Comet Klinkenburg-de Chéseaux, or more often Comet de Chéseaux, the relatively unknown amateur de Munck failed to get the credit he deserved. According to de Chéseaux, the comet initially lacked a tail, and instead resembled a nebulous star of the third magnitude with a coma some 5 arcminutes in diameter.

Beyond the orbit of Mars when first sighted, the comet brightened steadily as it approached the Sun. Attaining its maximum brightness of around magnitude −3 (as bright as Venus) on February 20, 1744, it displayed two tails; closest approach to Earth at 0.83 au (124 million km) took place on February 27, when it was only 12° from the Sun but bright enough to be glimpsed during the daytime, followed 2 days later by a perihelion of 0.22 au (33 million km).

Following perihelion the comet developed a spectacular tail that extended high into the eastern morning skies while its head lay beneath the horizon. Some 90° long, the tail consisted of six separate curving streams arrayed in a graceful fan-shape, the likes of which have been historically very rare. It is thought that the multiple tails were due to a large rotating nucleus, active in several areas and giving rise to vigorous jetting of material when exposed to sunlight. One witness—the 14 year old Sophie von Anhalt-Zerbst-Dornburg, later to become Catherine the Great—commented: "I had never seen anything so grand. It seemed very close to earth." Another witness of the same tender and impressionable age was Charles Messier, whose views of the comet must have had a profound effect on his choice of future occupation.

On March 9 de Cheseaux was the last to view the comet from Europe, but reports of it were forthcoming from Asia and the southern hemisphere as the comet headed south; its tail was reportedly 90° long on March 18. Incredibly, some Chinese observations contain reports of sounds being associated with the comet. This is not as far-fetched as it may sound, since crackling, hissing and whispering sounds have long been associated with aurorae, but only a select sensitive few can actually hear them. Like auroral sounds, the comet's audio track may have been produced by the interaction of particles with Earth's magnetosphere. Having been a naked eye object for no fewer than 110 days, C/1743 X1 was last sighted on April 22 (Figs. 4.10, 4.11, and 4.12).

C/1769 P1 (Messier)

Charles Messier's sixth comet discovery, made on the morning of August 8, 1769, was to become a Great Comet, his finest comet find. Within the space of a week it had brightened from being a faint telescopic object a few arcminutes across to a naked-eye spectacle with a tail 6° in length. By August 24 Chinese astronomers had

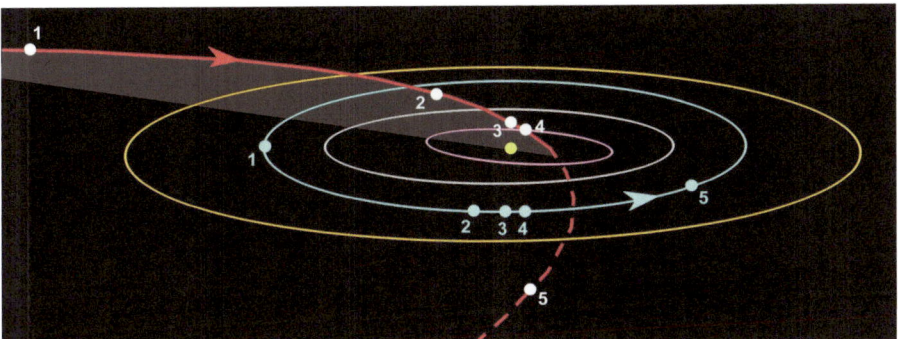

Fig. 4.10 The Great Comet of 1744, showing its path through the inner Solar System (*red line*: above ecliptic; *dashed red line*: below ecliptic) along with the orbit and positions of Earth (*blue*), and the orbits of Mars (*orange*), Venus (*gray*) and Mercury (*pink*). Key: (1) November 29, 1743; (2) February 20, 1744; (3) February 27; (4) March 1; (5) April 22 (Illustration © by the author)

Fig. 4.11 An impression by Hope Wolf of the Great Comet of 1744 over St. Martin in the Fields, London, on January 26

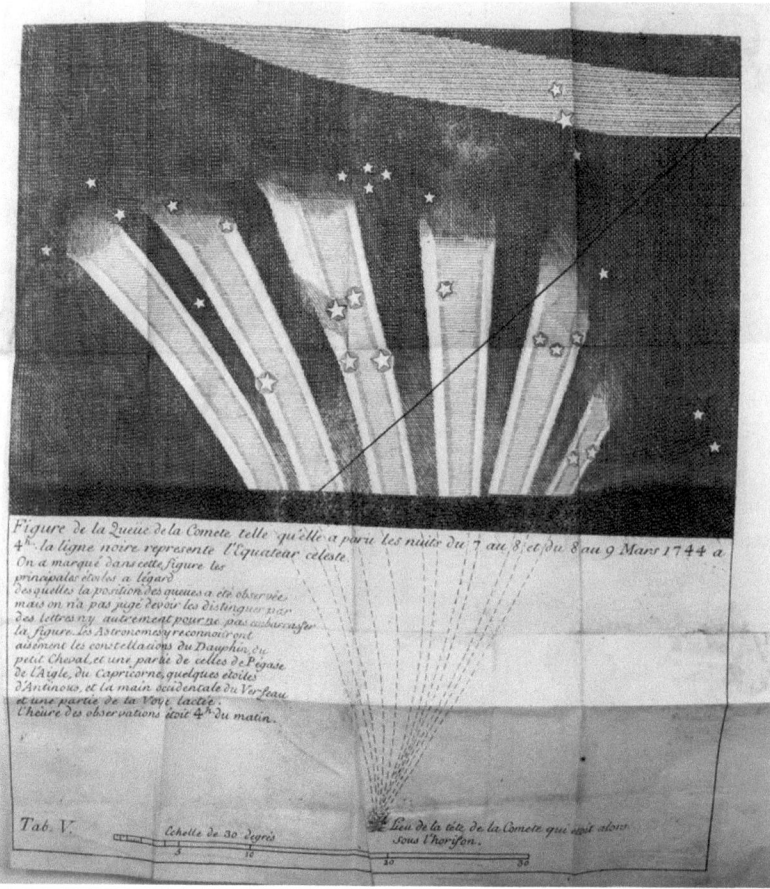

Fig. 4.12 The Great Comet at 4 a.m. on March 9, 1744, as seen from Paris. The tail projects 25° above the eastern horizon, almost reaching Delphinus, while the comet's head is 24° below the horizon. The curving tails have been drawn below the horizon to converge on the coma, but in reality the tails were probably due to active jets on a rotating nucleus, the tails being more parallel than converging to a point (Illustration from de Cheseaux' *Traité de la comète qui a paru en Décembre 1743 & en Janvier, Fevrier & Mars 1744*)

recorded it as a 'broom star.' On August 28, Messier gauged the comet's tail to be 15° long, but only 2 days later Captain James Cook, aboard the *Endeavour* in the South Pacific, estimated that its tail stretched for 42°. At the same time, astronomers Giacomo Maraldi and Jacques Cassini noted that the comet's tail appeared slightly curved, but from beneath slightly less transparent skies than Cook, estimated its tail to be 18° in length.

As its brightness increased, the comet's tail grew. Between September 3 and 5, Messier recorded that it grew from 36 to 43° long, while noticing the curvature of the tail (convex to the north) and the distinctly ruddy hue of the nucleus. Messier

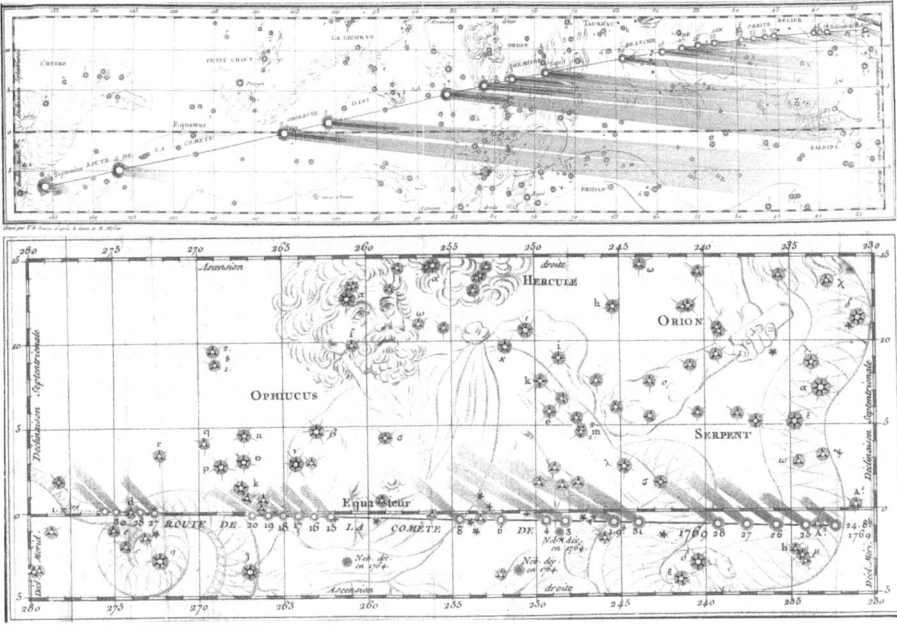

Fig. 4.13 Messier's charts showing the path of the Great Comet of 1769, from its discovery on August 8 to the end of the year published in his Grande Comète (1808)

also noted that the tail faded considerably at its extremity, and seemed to be composed of numerous parallel rays. On September 10 the comet made its closest approach to Earth, some 0.32 au (48 million km) while in Monoceros; Messier estimated that it was 60° long, while Alexandre Pingré and de la Nux, observing from a vessel in the Atlantic, thought that it approached 100° in length.

As perihelion approached and the comet was enveloped in the twilight, observations became more difficult, and the tail's length (which was always much brighter nearer the nucleus) appeared to contract. Maximum brightness occurred on September 22 when it shone at zero magnitude, but the visible tail was just a few degrees long. Making its closest approach to the Sun on October 8, 0.12 au (18 million km), the comet was picked up on October 23 from the Greenwich Observatory, after astronomers had determined its orbit and calculated its post-perihelion path. Messier alighted upon the comet on the following day using the same telescope with which he had discovered it, noting that it was just visible with the naked eye, shining at the fourth magnitude and describing it to have "...a bright nucleus, whose light equals that of the star, not clearly defined and surrounded by nebulosity, the tail is very weak and has a length of about 2 degrees." In early November, despite its fading nucleus, the tail stretched to some 6°, passing close to the globular clusters M10 and M12 in Ophiuchus. Messier's final observation of the comet was on December 1. In all, the Great Comet of 1769 had been visible to the unaided eye for 94 days (Figs. 4.13 and 4.14).

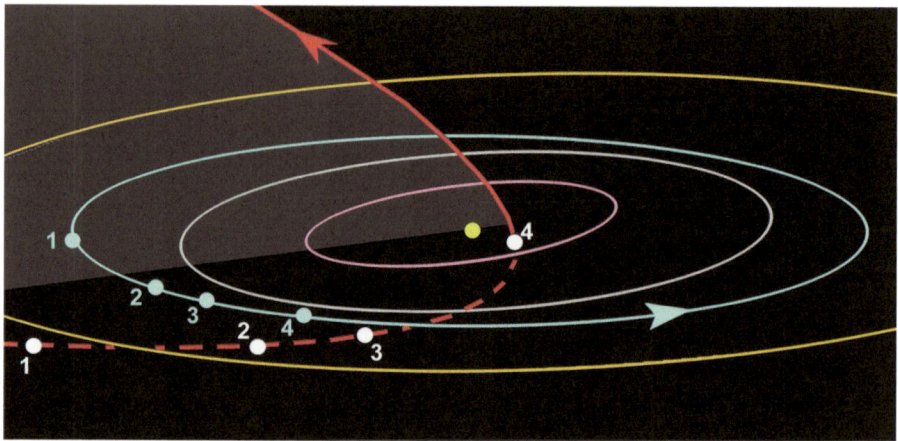

Fig. 4.14 The Great Comet of 1769, showing its path through the inner Solar System (*red line*: above ecliptic; *dashed red line*: below ecliptic) along with the orbit and positions of Earth (*blue*), and the orbits of Mars (*orange*), Venus (*gray*) and Mercury (*pink*). Key: (1) August 8; (2) September 10; (3) September 22; (4) October 8 (Illustration © by the author)

Here are Messier's notes on the Great Comet of 1769:

Great comet, discovered with a refractor on August 8 [1769], near the two stars of 9th magnitude above the head of Cetus, below and close to the ecliptic; observed in both branches of its orbit, 21 days in the first branch [before perihelion], and 21 days in the second branch [after perihelion], where it ceased its apparition on the first of December near the node of the tail of Serpens; in its first branch, it passed below Aries, through the feet of Taurus, traversed Orion, Monoceros, the equator and entered in the solar rays near Hydra, below the node. It re-appeared in the morning, when it left the rays of the Sun, on October 24 near Serpens, which it traversed as well as Ophiuchus, the equator, and ceased its apparition near the node of the tail of Serpens above and very close to the equator.

This comet has been one of the greater that have been observed. Its tail, in the night of September 9 to 10, had a length of 60 degrees; M. Pingre, who was in the sea, between Teneriffe and Cadiz, found it on September 11 of 90 degrees extent; and M. de la Nux, at the island of Bourbon, measured it in the morning of September 11 and found it at 97 degrees. Many calculations have been made of this Great Comet by Leonard Euler and Lexell, for finding its true elliptical orbit and its period."

Chapter 5

Victorian Spectacles: Great Comets of the Nineteenth Century

Nature couldn't have chosen a better century upon which to bestow such an unprecedented profusion of brilliant comets. During the nineteenth century, the mathematics involved in pinning down cometary orbits was pretty well established. The instrumentation with which to study comets had never been more advanced, and the sheer number of observers—both amateur and professional—interested in the study of comets was at an all-time high. No fewer than eight Great Comets graced nineteenth-century skies, six of these appearing within the space of just 40 years, bestowing upon many an astronomer an enviable set of indelible memories.

Great Comet C/1807 R1

When first seen from Sicily by Castro Giovanni in the evening twilight of September 9, 1807, comet C/1807 R1 was already a naked-eye object sporting a short tail. It must have presented a memorable sight, lying in Virgo close to the star Spica, with Venus, Mars, and Saturn all located nearby while the first quarter Moon and Jupiter lay in the south.

Moving northeast and among the planets arrayed near the western horizon, the comet underwent perihelion on September 19, 0.65 au (97 million km) from the Sun, and the following day had attained a maximum brightness of the first magnitude. It was only at this time that the comet came to the attention of many astronomers, including Honoré Flaugergues and Jean-Louis Pons, both of whom claimed co-discovery.

At its perigee of 1.15 au (172 million km) from Earth on September 27, the comet had developed a curving tail some 8° long. Moving through Serpens in

P. Grego, *Blazing a Ghostly Trail: ISON and Great Comets of the Past and Future*, The Patrick Moore Practical Astronomy Series, DOI 10.1007/978-3-319-01775-4_5, © Springer International Publishing Switzerland 2014

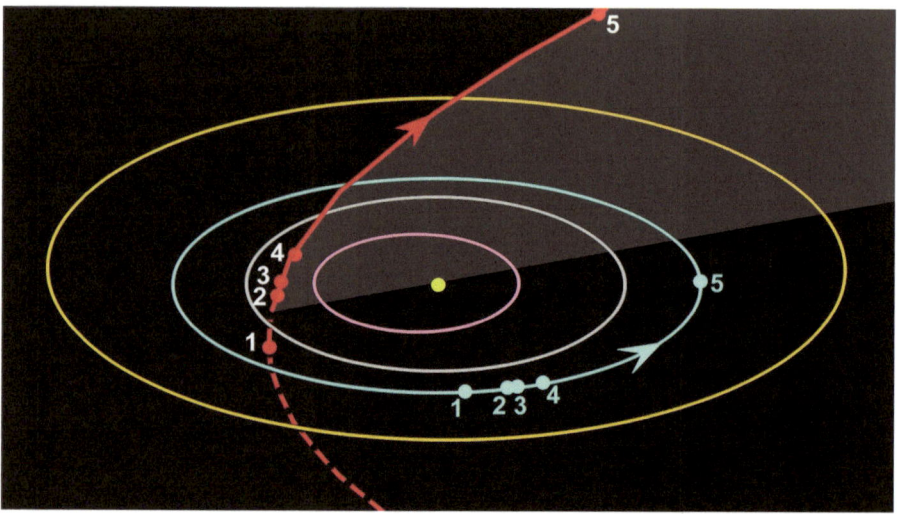

Fig. 5.1 The Great Comet of 1807, showing its path through the inner Solar System (*red line*: above ecliptic; *dashed red line*: below ecliptic) along with the orbit and positions of Earth (*blue*), and the orbits of Mars (*orange*), Venus (*grey*) and Mercury (*pink*). Key: (1) September 9, 1807; (2) September 19; (3) September 20; (4) September 27; (5) December 8 (Illustration © by the author)

mid-October, still shining at the first magnitude in the evening sky, the comet displayed two tails, the straighter and longer of which was some 10° in length. Crossing Hercules in late October and early November the comet faded to the fourth magnitude, but its main tail retained a length of 5°. On November 20 William Herschel measured its tail at 2½ degrees long. Having been visible with the unaided eye for 90 days, the Great Comet of 1807 faded from view after the first week of December as it passed through Cygnus, passing near Deneb. It was last seen telescopically by Vincent Wisniewski at St Petersburg, Russia, on March 27, 1808 (Fig. 5.1).

Among the comet's many followers was an elderly Charles Messier, who wrote:

This comet became very beautiful, and stayed beautiful during a large number of days and was seen all around the world; it was marked in the sky by a nucleus of great luminosity which it enveloped, and from which came out a very clear, very extended tail, which could have 4 or 5 degrees or more and which let see through it many small [faint] stars. It is said…that this comet was the most beautiful which has appeared since sixty years, one had to say since 38 years, which is since that which appeared in 1769, the greatest which has been observed, which had a tail of 97 degrees and which fixed the birth of Napoleon…

This beautiful comet of 1807 which attracted the views of the public in the evenings, also all the night, united the world for the view. My eyesight became still fainter and worse than in 1805 which took away the hope to observe it with care. Nevertheless, I investigated it with a night refractor and I marked its configuration with the neighboring stars from

October 19 to the 26th of the same month. Afterwards, I observed it more exactly with my great achromatic refractor, equipped with its micrometer, from November 3 until January 26, 1808. At these observations, I have made use of help for the first time during all my observations by a person who counted at the clock and marked the divisions of the instrument which my eyes could no more perceive…

The nucleus of this beautiful comet has presented to some astronomers an observation worth seeing and publishing, so here is it: I observed on November 3 and 9 with a [reflecting] telescope of two meters length and have seen the nucleus, round and of uniform light, I have pain to believe in this observation…In all the comets I have observed, I have always seen the nucleus of Great Comets surrounded by a more or less vivid light, without being terminated…I am well of the opinion…that the nucleus, or the diameter of the solid matter of comets is of an extreme smallness and almost impossible to see a diameter terminated by a round circle and of equal light.

Great Comet C/1811 F1

One of history's most celebrated comets, the Great Comet of 1811, was visible with the unaided eye for 260 days—a record that it held until C/1995 O1 (Hale-Bopp), which was visible to the naked eye for 569 days. Discovered by Flaugergues at Viviers, southern France, on March 25, 1811, the comet lay in the southern constellation of Puppis (then part of Argo Navis), coming in from the south on a plane almost perpendicular to the ecliptic. An evening object, it had brightened to naked-eye visibility by April 11, and slowly moving northwards it crossed Monoceros and Canis Minor during April and May, while its apparent angular distance from the Sun decreased. It became increasingly embedded in the evening twilight, shining at the fifth magnitude. From Paris, Alexander von Humboldt was the last to observe the comet (prior to its conjunction with the Sun) at dusk on June 16, when it was in Cancer, 40° east of the Sun.

It was recovered again during the third week of August while in Leo Minor, sufficiently north of the Sun to be seen in the dusk and dawn skies, shining at the second magnitude. During the autumn of 1811 the comet became a really conspicuous sight in the northern skies. Perihelion at 1.04 au (156 million km) took place on September 12, the comet shining as a first magnitude object with a coma larger than the Moon's apparent size and a tail around 12° long. The comet developed two distinct streams that flowed from the coma's edge and merged further along the tail. Visible to European and American astronomers throughout the night for many weeks, it passed through Ursa Major, Canes Venatici and Boötes, reaching its most northerly declination on October 3, when its tail extended 25° and measured 6° wide at its broadest.

William Herschel made paid great attention to the comet. He noted its well-defined nucleus [false nucleus] and measured it to be 689 km across; it appeared to be of a 'ruddy hue' surrounded by a bluish-green nebulosity. The comet's head was measured to be 204,000 km across, enveloped in a coma some 1.35 million km wide. While the apparent length of its tail was shorter than many other comets, its distance and angle with respect to Earth actually meant that the tail's true length

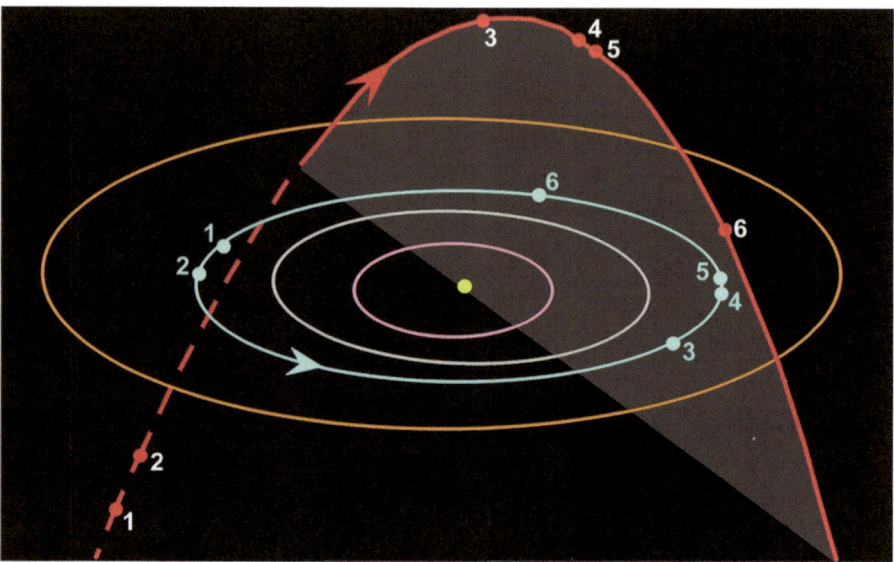

Fig. 5.2 The Great Comet of 1811, showing its path through the inner Solar System (*red line*: above ecliptic; *dashed red line*: below ecliptic) along with the orbit and positions of Earth (*blue*), and the orbits of Mars (*orange*), Venus (*grey*) and Mercury (*pink*). Key: (1) March 25, 1811; (2) April 11; (3) September 12; (4) October 16; (5) October 20; (6) January 1, 1812 (Illustration © by the author)

reached 193 million km—greater than the Earth-Sun distance—and that the tail was 24 million km across at its broadest. It was noted that the comet had such a large perihelion distance—again, greater than the distance between Earth and the Sun—that it never really showed what it was truly capable of. The comet was grand enough, however, to be considered by Napoleon Bonaparte as being a good omen for his planned 1812 invasion of Russia.

Having made its closest approach to Earth on October 16, a perigee of 1.22 au (183 million km), the comet's maximum brightness of magnitude 0 was observed four days later. Traversing Boötes and Hercules as an evening object during late October, the comet faded by a magnitude but retained a more than 20° long tail. By November 13, Herschel was no longer able to detect the nucleus but noted that the two streams bounding the brightest part of the 7° long tail were 4 and 3½ degrees in length. In early December the comet, now shining at the third magnitude and having a 5° tail, passed near Altair in Aquila. By early January 1812 the Great Comet was in Aquarius, just visible with the unaided eye and immersed in the evening twilight. It was followed telescopically until August 17, 1812, by Wisniewski at Neu-Tscherkask, Russia, having been the last to observe it, who estimated that it was a barely detectable smudge of the 12th magnitude (Figs. 5.2, 5.3, and 5.4).

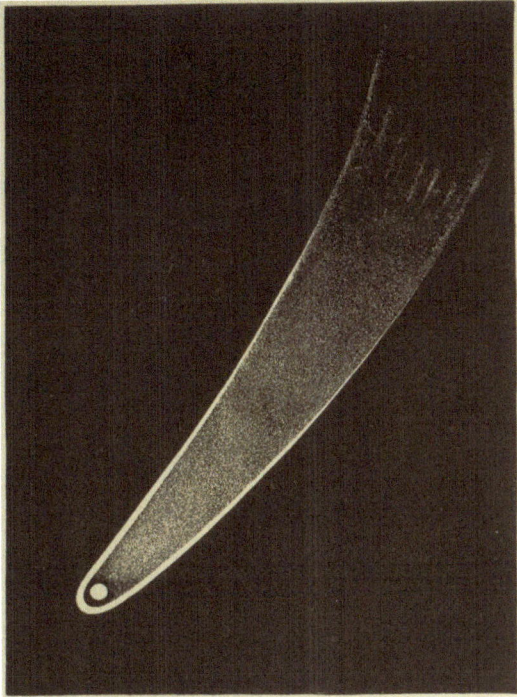

Fig. 5.3 The Great Comet of 1811, from Daniel Kirkwood's *Comets and Meteors* (1873)

Great March Comet C/1843 D1

More than 30 years elapsed before the next Great Comet appeared. While the Great Comet of 1811 had taken a somewhat leisurely path through the inner Solar System and was observed for 18 months, the Great March Comet of 1843 streaked through the skies at a blistering pace. Arriving at, and leaving, the Sun's vicinity south of the ecliptic, it only nudged north of the ecliptic for a day or so near perihelion. Discovered in Cetus on February 5, 1843, the Great March Comet sped rapidly towards an extremely close perihelion of just 0.006 au (898,000 km), less than a solar diameter away from the Sun, on February 27. Shining at magnitude −3 it was clearly visible in daylight at this time, less than a degree from the Sun.

Perigee took place on March 6, when the comet passed 0.84 au (126 million km) from Earth, on the following day reaching its maximum brilliance of magnitude −6. C/1843 D1 belongs to the family of comets known as Kreutz Sungrazers, members of which were produced around AD 1106 when the parent comet, X/1106 C1 (a widely observed Great Comet), broke up into multiple fragments. As their name suggests, Kreutz Sungrazers pass extremely close to the Sun at perihelion and are liable to become extremely brilliant.

Fig. 5.4 Impression of the Great Comet of 1811 as seen in the dusky skies of October 3 (Illustration © by the author)

On March 7, a New Haven, Connecticut, observer described the tail as "a long, narrow, and brilliant beam, slightly convex upwards, the lower end being apparently below the horizon."

C/1843 D1 developed a tail of more than 2 au (300 million km)—beating the tail length record set by C/1811 F1—although by this time it was more favorably viewed from the southern hemisphere. Estimates of the tail's apparent length in early and mid-March varied considerably between observers, ranging between 25 and 45°.

People in the southern hemisphere were enthralled by the comet. A letter published in *Hobart Colonial Times* (Tasmania) reads: "There is a great doubt in the public mind as to this phenomenon, and many people will not believe that this is a comet…[I]f a comet were so close to the earth as this meteor evidently is we should stand a good chance of being well-roasted."

From Port Macquarie, Australia, Annabella Innes recorded the comet in her journal: "…we saw it first on Friday, 3rd, at about six o'clock…looking due west over the lake we beheld a really splendid comet."

Fig. 5.5 Astronomer Charles Piazzi Smyth's painting of the Great March Comet of 1843, as seen from the Cape of Good Hope, South Africa

On March 10 she wrote:

Extremely warm all day. We hear that some of the people at Port Macquarie firmly believe that the world is coming to an end, and say it was prophesied a hundred years ago that a comet would appear in 1843 which would destroy the world. This sounds very alarming, but we heard nothing about it till the comet appeared, and now it is easy for everyone to prophesy. We half hoped the comet would not be seen tonight, for though it is very beautiful and we have a certain amount of pleasure in looking at it, we have also an indescribable dread of it.

As it headed south the comet faded, and by March 26 it was no longer visible with the unaided eye. It was last observed on April 19 by Thomas Maclear at the Cape of Good Hope (Figs. 5.5 and 5.6).

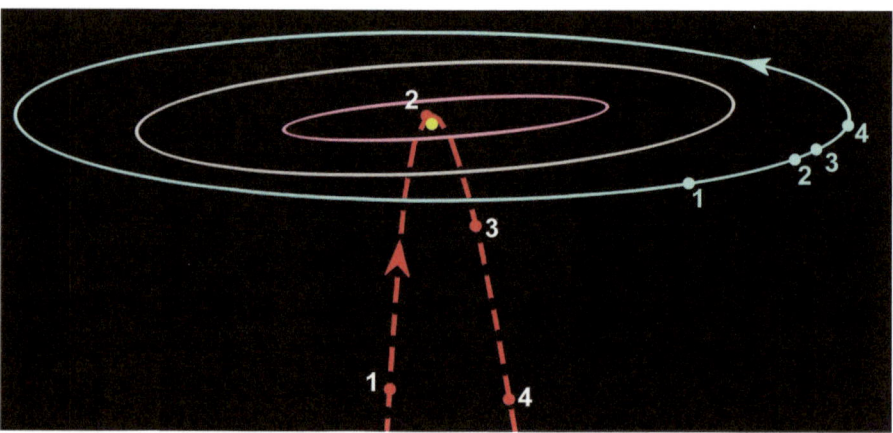

Fig. 5.6 The Great March Comet of 1843, showing its path through the inner Solar System (*red line*: above ecliptic; *dashed red line*: below ecliptic) along with the orbit and positions of Earth (*blue*), and the orbits of Venus (*gray*) and Mercury (*pink*). Key: (1) February 5; (2) February 27; (3) March 6; (4); March 26 (Illustration © by the author)

C/1858 L1 (Donati)

At 10 p. m. on June 2, 1858, Giovanni Donati at Florence, Italy, discovered a faint comet in Leo. It brightened to become one of the most brilliant and most celebrated of the nineteenth century. "From being a comparatively obscure observer, Donati found himself suddenly the astronomical hero of the day, for his brilliant comet not only formed an interesting subject for intelligent study but it also created for a time a lively taste for astronomy among all classes of the community." (*Monthly Notices of the RAS*, 34, 153, 1874). In *Curiosities of the Sky* (1909), Garrett Serviss wrote:

> The splendid comet of 1858, usually called Donati's…was, perhaps, both as seen by the naked eye and with the telescope, the most beautiful comet of which we have any record. It too marked a rich vintage year, still remembered in the vineyards of France, where there is a popular belief that a great comet ripens the grape and imparts to the wine a flavor not attainable by the mere skill of the cultivator. There are 'comet wines,' carefully treasured in certain cellars, and brought forth only when their owner wishes to treat his guests to a sip from paradise.

Leaving aside the myth of 'comet wines' there seems no doubt that Donati's Comet was an incredibly beautiful sight that left a deep and lasting impression on its viewers. Inspiring art and poetry, Alexander J. D. D'Orsey penned a verse entitled "The Great Comet of 1858":

> Then came the climax! Oh that glorious hour!
> The mighty Comet in its pride of power!
> No sight like that had ever met my gaze!
> No sight like that will living man amaze!
> Beautiful vision! Feathery, graceful, bright,
> A starry diamond in a veil of light!

Fig. 5.7 Impression of Donati's Comet (over an imaginary city) by E. Weiss, as it would have appeared on October 5, 1858

On August 28, K. C. Bruhns at Berlin was the first to detect the comet with the unaided eye, and it remained a naked-eye object for 80 days. It is known that one of the comet's observers on September 14 was a whiskerless, 49-year-old prairie lawyer and Senatorial candidate named Abraham Lincoln, who sat on the porch of his hotel in Jonesboro, Illinois, to see 'Donti's Comet' in the evening skies. On September 27, William Usherwood from Walton-on-the-Hill, England, made history by taking the first ever comet photograph, imaging Donati's Comet using a simple portrait camera; this photograph is now lost, but other pioneering photographers managed to secure pictures of the comet in the weeks that followed.

Perihelion of 0.58 au (87 million km) took place on September 30, when it shone at first magnitude and sported a tail some 30° long. A week later it had reached its maximum brightness of magnitude 0, and its tail stretched 40°. Perigee occurred on October 11, the comet passing 0.54 au (81 million km) from Earth. By this time it had developed a prominent dust tail some 60° long, curving like a gigantic scimitar, along with two long very thin, straight, ion tails. Great structure around the false nucleus and in the coma, consisting of sharply defined shells and arcs of material ejected from the active rotating nucleus, was a notable hallmark of Donati's Comet, and this repaid very close, high magnification scrutiny through large instruments.

Donati's Comet was last seen on March 4, 1859, by C. W. Moesta at the National Observatory of Santiago, Chile (Figs. 5.7, 5.8, 5.9, 5.10, and 5.11).

Fig. 5.8 The intricate head of Donati's Comet, as observed by George Bond with the 15-in. Merz refractor of Harvard College on October 2, 1858. Bond's description: "The nucleus…was unusually bright, and rounded on the side toward the Sun. An increase of brilliancy in the nucleus was afterwards recognized as the precursor of a fresh eruption from its surface…There were three dark openings in the innermost envelope, between which it was intersected with bright rays"

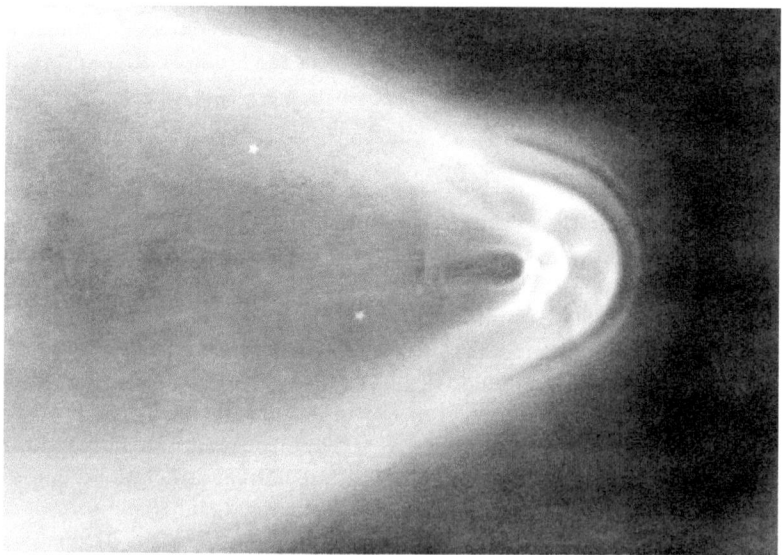

Fig. 5.9 The intricate head of Donati's Comet, as observed by H. G. Fette with the 15-in. Merz refractor of Harvard College on October 10, 1858

Fig. 5.10 William Turner's impression of Donati's Comet over Oxford

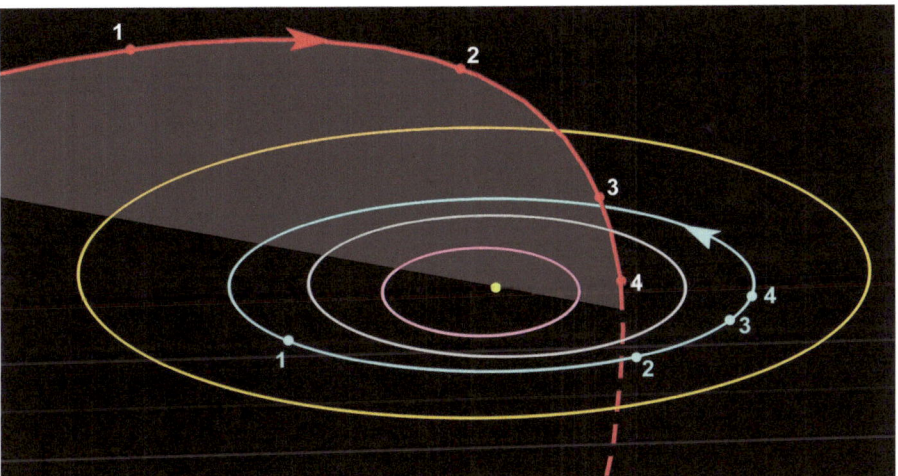

Fig. 5.11 Donati's Comet, showing its path through the inner Solar System (*red line*: above ecliptic; *dashed red line*: below ecliptic) along with the orbit and positions of Earth (*blue*), and the orbits of Mars (*orange*), Venus (*gray*) and Mercury (*pink*). Key: (1) June 2, 1858; (2) August 28; (3) September 30; (4) October 11. (Illustration © by the author)

Great Comet C/1861 J1

Perigee: 1861/Jun/30 0.13
  Max brightness: 1861/Jun/27 0
  Discovered by John Tebbutt, observing from Windsor, New South Wales, Australia, on May 13, 1861, the comet was already shining at magnitude of +4 and easily visible with the unaided eye. It remained visible only from the southern hemisphere through its perihelion on June 12, when it passed 0.82 au (123 million km) from the Sun until it was accessible to northern hemisphere viewers from June 29 onwards. At this time it was an apparent angular distance of 11½ degrees from the Sun. Perigee took place on June 30, the comet passing 0.13 au (19 million km) from Earth, at which time the comet shone at magnitude −2 and had a tail exceeding 90° in length. For 2 days at perigee, Earth actually lay within the comet's tail, and it is reported that streams of cometary material radiating from the nucleus could be discerned.
  In his journal entry of July 1, Granville Stuart in western Montana wrote: "Saw a huge comet last night in the northwest. Its tail reached half across the heavens. It has probably been visible for some time, but as it has been cloudy lately I had not observed it before."
  A journal entry from Sarah Epsy in Alabama on the same day reads: "A brilliant and beautiful comet appeared tonight in the same part of the heavens as that a few years ago, the train of this is the longest that I ever saw, pointing directly upwards."
  On July 5, James Robinson aboard the schooner Conchita in Agiabampo Harbor, Mexico, noted: "I awoke in the night at 1 o'clock, when I had a glorious sight of the largest comet I ever beheld. The head, or nucleus, was large as Venus, and very bright and blazing, and about 20° above the horizon, pointed to the north, while the bright, long tail reached full half way across the heavens. It was a most wonderful sight."
  The Great Comet of 1861 was visible to the naked eye for around 90 days, from the date of its discovery to August 15, and was followed through telescopes until May 1862 (Figs. 5.12 and 5.13).

Great Southern Comet C/1865 B1

This was an almost exclusively southern hemisphere comet, widely observed from Australia, New Zealand, South Africa and South America. Perihelion took place on January 14 at 0.03 au (4.5 million km) from the Sun, with perigee 2 days later, 0.94 au (141 million km) from Earth. First visible with the unaided eye on January 17, C/1865 B1 remained a naked-eye object for 36 days. The comet was at its brightest on January 24, a first magnitude object. It was last seen telescopically on May 2 (Fig. 5.14).

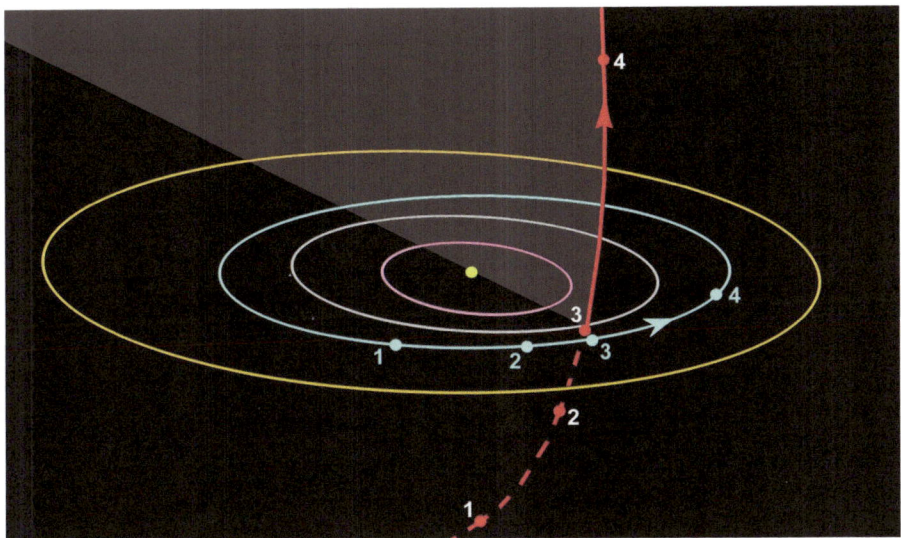

Fig. 5.12 The Great Comet of 1861, showing its path through the inner Solar System (*red line*: above ecliptic; *dashed red line*: below ecliptic) along with the orbit and positions of Earth (*blue*), and the orbits of Mars (*orange*), Venus (*gray*) and Mercury (*pink*). Key: (1) May 13; (2) June 12; (3) June 27; (4) August 15 (Illustration © by the author)

Fig. 5.13 Impression of the Great Comet of 1861 by E. Weiss

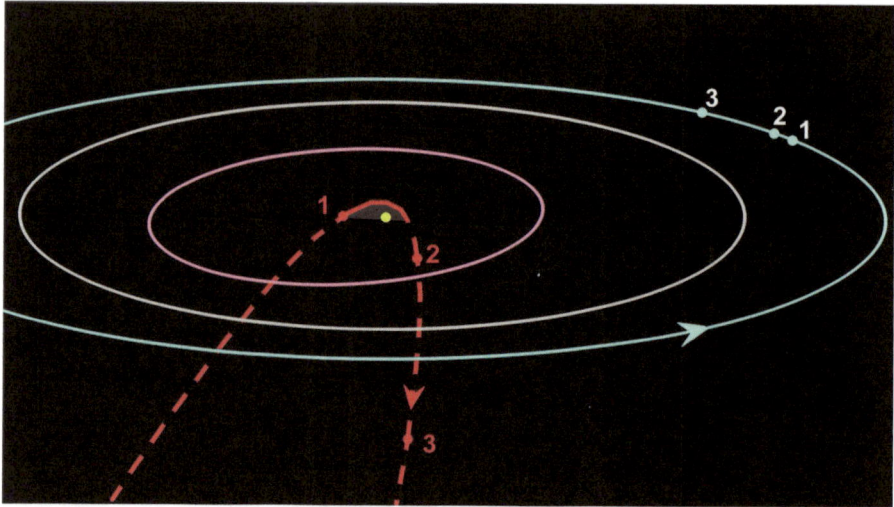

Fig. 5.14 The Great Southern Comet of 1865, showing its path through the inner Solar System (*red line*: above ecliptic; *dashed red line*: below ecliptic) along with the orbit and positions of Earth (*blue*), and the orbits of Venus (*grey*) and Mercury (*pink*). Key: (1) January 14; (2) January 16; (3) January 24 (Illustration © by the author)

C/1874 H1 (Coggia)

Discovered in Camelopardalis on April 17 by Jérôme Coggia at Marseilles, C/1874 H1 became visible without optical aid on June 10 and remained a naked-eye object for 50 days. Perihelion took place on July 9, passing 0.68 au (102 million km) from the Sun, at which time William Denning estimated the tail to be 10° in length. Four days later it reached its peak brightness of magnitude −1. In his journal entry for July 13 the poet Gerard Manley Hopkins made the following entry: "The comet − I have seen it at bedtime in the west, with head to the ground, white, a soft well-shaped tail, not big: I felt a certain awe and instress (Peter, not a mistake?), a feeling of strangeness, flight (it hangs like a shuttlecock at the height, before it falls), and of threatening."

On July 23 Coggia's Comet approached Earth within 0.29 au (44 million km), when its tail was reported to stretch for 60°. Like Donati's Comet, Coggia's Comet displayed intricate structure around the false nucleus and coma, and the changing appearance in the head of the comet between June 10 and July 14 was closely monitored.

Famously, the comet's nucleus was the first to undergo a thorough spectroscopic analysis by pioneers William Huggins, Angelo Secchi, Norman Lockyer, Georges Rayet and Charles Wolf, who concurred that the comet's light in the continuous spectrum was reflected sunlight and that it was polarized. There was some disagreement over the bright bands visible in the spectrum—carbon was detected, but Secchi was unable to verify the existence of hydrocarbons (Figs. 5.15 and 5.16.

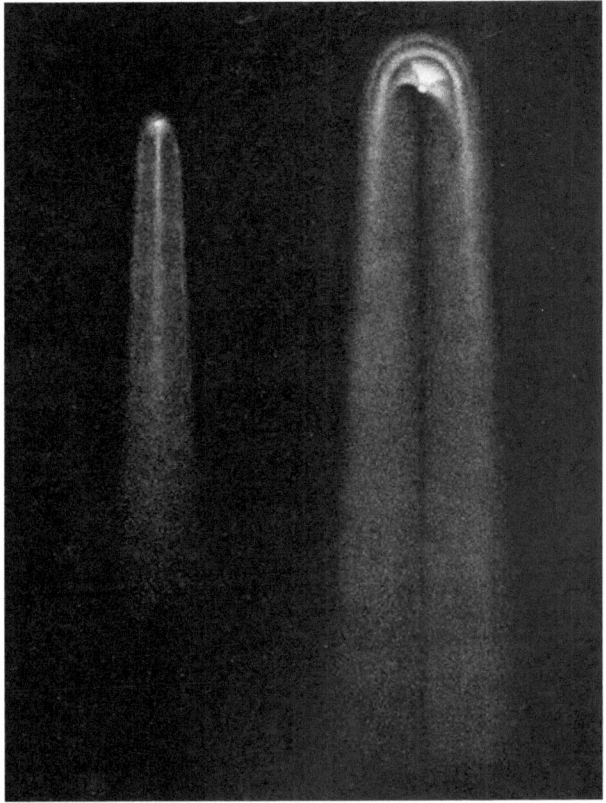

Fig. 5.15 Coggia's Comet, observed on June 10 and July 9 (From Robert Ball's *The Story of the Heavens*, 1893)

Great September Comet C/1882 R1

First discernible with the naked eye from the southern hemisphere on September 1, the Great September Comet of 1882 remained visible without optical aid for 135 days. On the morning of September 7, William Finlay at the Royal Observatory in Cape Town was the first astronomer to properly observe and record the comet, noting that it was a third magnitude object with a tail 1° long. Brightening extremely rapidly as it sped towards the Sun, it shone at the first magnitude a week later. On September 16 it passed 0.99 au from Earth, a day later reaching a perihelion of 0.008 au (1.2 million km). For a period of less than a day at around perihelion, C/1882 R1 was easily visible in the daytime in close proximity to the Sun, shining at a magnificent magnitude of −17. Like the Great March Comet C/1843 D1, C/1882 R1 was a member of the Kreutz Sungrazer family.

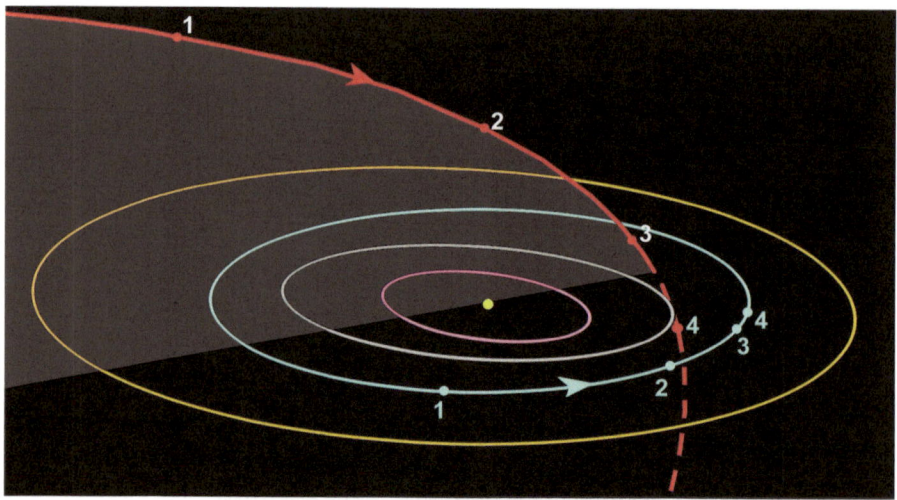

Fig. 5.16 Coggia's Comet, showing its path through the inner Solar System (*red line*: above ecliptic; *dashed red line*: below ecliptic) along with the orbit and positions of Earth (*blue*), and the orbits of Mars (*orange*), Venus (*gray*) and Mercury (*pink*). Key: (1) April 17; (2) June 10; (3) July 9; (4) July 23 (Illustration © by the author)

Following perihelion the comet actually transited the face of the Sun. Observed the comet with a neutral density filter (keeping the Sun's light manageably dim while retaining a view of the comet), Finlay observed C/1882 R1 to first contact with the Sun's edge, but as soon as the transit commenced not a trace of it could be discerned against the solar surface.

Entering darker skies, again favorably visible from the southern hemisphere, the comet faded only very gradually. Intriguingly, on September 30 there were reports (including from Finlay and the eagle-eyed Edward Barnard) that the comet's nucleus had elongated and then split into two bright components. No fewer than five fragments were noted on October 17, each of which varied in brightness from night to night.

By mid-October the comet had developed an anti-tail pointing in the Sun's direction; such phenomena are illusory, the result of viewing a comet's curving tail from a certain angle. The comet remained visible without optical aid until February 6, 1883, and the last telescopic sighting was made by B. A. Gould at Córdoba, Argentina, on June 1, 1883 (Figs. 5.17 and 5.18).

Fig. 5.17 Photograph of C/1882 R1 by David Gill (Cape Town)

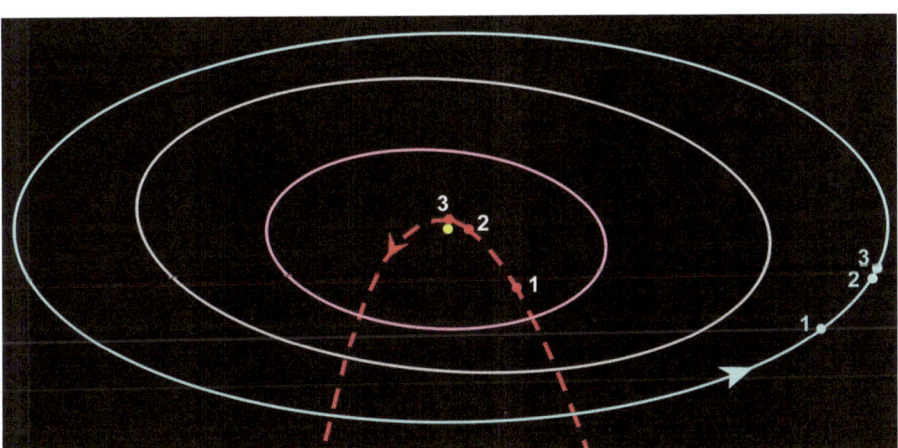

Fig. 5.18 The Great September Comet of 1882, showing its path through the inner Solar System (*red line*: above ecliptic; *dashed red line*: below ecliptic) along with the orbit and positions of Earth (*blue*), and the orbits of Venus (*gray*) and Mercury (*pink*). Key: (1) September 1; (2) September 16; (3) September 17 (Illustration © by the author)

Chapter 6

Comets in Plain View: Great Comets of the Twentieth Century

Two waves of Great Comets were to sweep across twentieth-century skies. The four decades separating these waves saw tremendous advances in science and astronomy, and humanity finally got to see comets up-close through the eyes of space probes.

Great Comet C/1901G1

When first seen in the morning skies of April 12, 1901, C/1901G1 was already of naked-eye magnitude and displayed a short tail. An official named Viscara of Paysandu, Uruguay, was recognized as the first to see and report the comet to the astronomical community, so the comet is sometimes referred to as Comet Viscara. An object almost exclusively viewed from the southern hemisphere, this first Great Comet of the twentieth century rapidly brightened as it approached perihelion on April 24, some 0.24 au (35 million kilometers) from the Sun. Robert Innes, observing from Cape Town, reported:

> …[on April 24] when day was breaking, I had begun to despair of seeing any comet, but on giving a final look round in very bright twilight I saw two shafts of light rising above the mountains in the east. A few minutes later the comet had entirely risen. It was a brilliant object with a bright nucleus and a tail about 10° in length, curved on the southern side. The colour of all was a very deep yellow, but the comet was very near the horizon. Through the 10-in. guiding telescope (now in broad daylight) the yellow tint of the nucleus was very marked. There was no coma visible, the tails (see drawing) springing directly from the nucleus. By comparison with Mercury, the nucleus was estimated to be about two thirds of Mercury's diameter, which makes it about 4 arcs; its brightness was about equal to Mercury's [Mercury was then shining at magnitude −0.4].

P. Grego, *Blazing a Ghostly Trail: ISON and Great Comets of the Past and Future*,
The Patrick Moore Practical Astronomy Series, DOI 10.1007/978-3-319-01775-4_6,
© Springer International Publishing Switzerland 2014

Its perigee of 0.83 au (124 million kilometers) from Earth took place on April 30, and at its brightest in the evening twilight skies on May 5 shone at magnitude −2 and had a 25° long multiple tail. On May 10, as the comet receded, Australia's *Alpine Observer* newspaper reported:

> The comet is still visible in the heavens after sunset. On Monday night, owing to the clearness of the sky, Mr. Barrachi was able to make a good observation of the triple-tailed visitor. It had a particularly brilliant appearance, even when looked at with the naked eye, and the Government astronomer states that it was brighter than ever since its discovery. The comet continues to travel in a north easterly direction, but at a slightly diminished pace.

Observing from Cape Town, Joseph Lunt described the comet's post-perihelion appearance:

> The most remarkable feature of the comet, viz. the long faint preceding tail, did not become visible until the comet had emerged from the strong twilight. It was first seen on the evening of Friday, May 3, as a faint ray, scarcely distinguishable, springing from the head at an angle of about 40° to the main tail. This faint tail appeared on two photographs taken with a portrait lens the same evening. On the two following nights, however, as the comet receded further from the Sun and became visible against a darker sky, it was a most conspicuous feature. On the evening of Monday, May 6, the faint tail was seen to be quite four times as long as the main tail and fully 30° in length, but fading away so gradually that it was difficult to place any exact limit on it. At this time the comet attained its maximum splendor as a naked-eye object. With an exposure of 25 min a portrait lens showed not only the main faint tail, but two still fainter rays between it and the bright tail…The space on each side of the faint rays was willed with faint light, and the darker space between them showed clearly by contrast…

In all, the Great Comet of 1901 was visible for 38 days without optical aid, fading from naked-eye view by May 20, but was telescopically followed through to October (Figs. 6.1, 6.2, 6.3, 6.4, and 6.5).

Great January Comet C/1910 A1

In early 1910, as astronomers around the world eagerly prepared for the predicted showing of Halley's Comet, a new comet was discovered which would prove to far outshine Halley itself. Like the Great Comet of 1901, it was first noticed from the southern hemisphere when it was already a naked-eye object; it is commonly accepted that it was first spotted by diamond miners in the Transvaal, South Africa, in the dawn twilight of January 12. Shining at magnitude −1 and yet to reach perihelion, it was clear that this comet was going to be pretty spectacular. From the Transvaal Observatory, Johannesburg, Robert Innes first began studying the comet on January 17, the date of its perihelion of 0.13 au (19 million kilometers) from the Sun, when it had become so bright—magnitude −4—that it could clearly be discerned during daylight without optical aid, just 4½° from the Sun. Perigee of 0.86 au (124 million kilometers) from Earth occurred just a day later.

After perihelion the Great Comet of 1910 became visible in the evening skies of the northern hemisphere. Due to a misheard telephone report from the southern

Fig. 6.1 Image of the Great Comet of 1901, taken with the McClean Telescope, Cape Town (exposure 15 min) on May 4

Fig. 6.2 Image of the Great Comet of 1901, taken with a portrait lens by Joseph Lunt on May 7

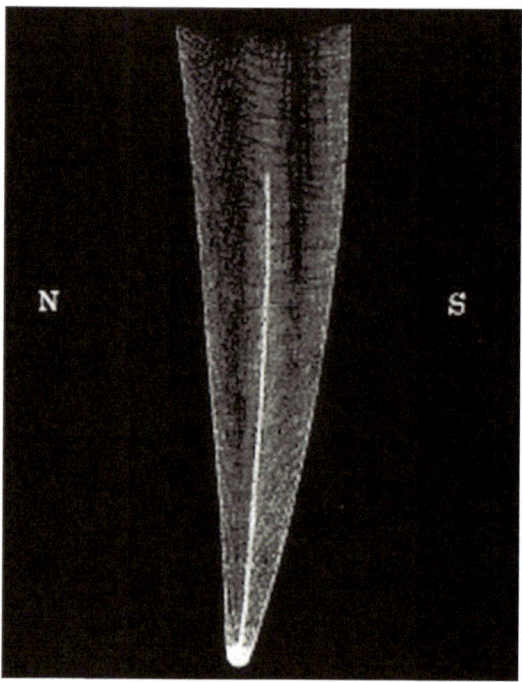

Fig. 6.3 Drawing of the Great Comet of 1901 at perihelion by Robert Innes from Cape Town, viewed in strong morning twilight on April 24

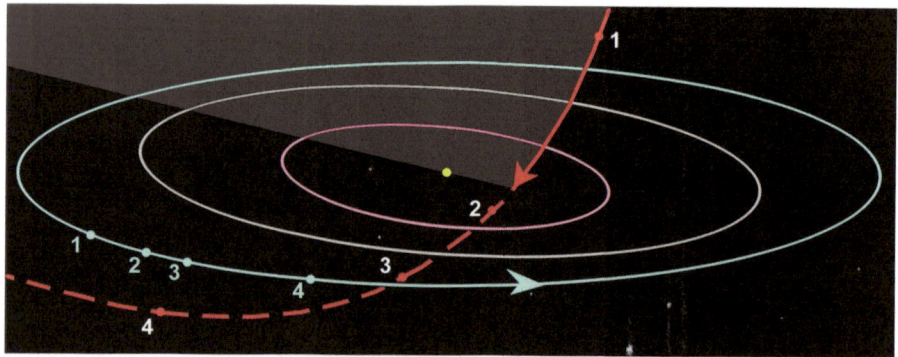

Fig. 6.4 The Great Comet of 1901, showing its path through the inner Solar System (*red line*: above ecliptic; *dashed red line*: below ecliptic) along with the orbit and positions of Earth (*blue*), and the orbits of Venus (*gray*) and Mercury (*pink*). Key: (1) April 12, 1901; (2) April 24; (3) April 30; (4) May 20. (Illustration © by the author)

Fig. 6.5 Impression of the Great Comet of 1901, based on a drawing by Lunt made at 7.15 p.m. on May 12. (Illustration © by the author)

hemisphere, the comet mistakenly became known as Drake's Comet (rather than Great Comet). Although fading, it developed a superb gently curving tail that extended to some 50° by the end of January, when it was located near the border of Aquarius and Pegasus and visible against an astronomically dark sky background. The comet's head appeared small and nebulous, shining at the 3rd magnitude. According to *The New York Times* of January 30:

> Its appearance is reported to have caused extreme terror among the Russian peasants, who regard it as the precursor either of a great war in the Far East or of the end of the world. Warnings have been issued as to the effect it is likely to have upon the populations of North Africa and India…Its tail is estimated to be 9,000,000 miles in length and its speed nearly 1,000,000 miles an hour. Mr. Crommelin of the Greenwich Observatory describes it as the brightest comet seen since 1888. It has a very conspicuous tail with a distinct dark streak running through the middle similar to the comet of 1874, while, as in the case of the great comet of 1878, its curvature is on a small scale…its orbit has been described by Sir Robert Ball as narrow and in the shape of a hairpin. The comet will soon cease to be visible…Its brightness was considerably diminished this week…Walter Maunder of the Greenwich Observatory thinks this particular comet has never been seen before, and may never be seen again, although observations are not exact enough to determine whether its course is elliptical or parabolic…

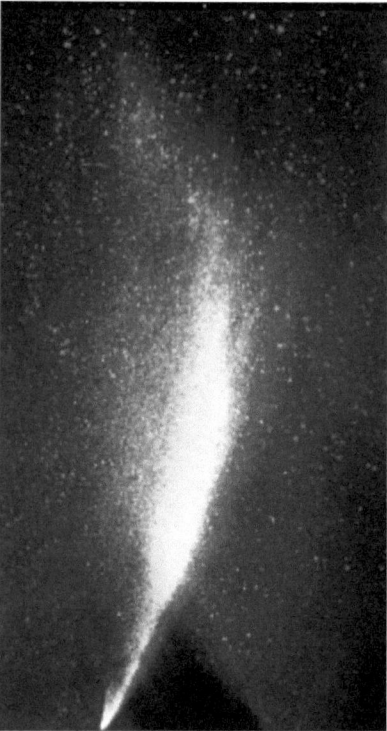

Fig. 6.6 Photograph of the Great Comet of 1910 by Carl Lampland and Vesto Slipher using the 24-in. refractor at the Lowell Observatory on January 28

By February 7 the comet's head had faded to the 4th magnitude, but it still showed a faint, ghostly tail some 20° long. After having been visible with the naked eye for some 26 days, the comet faded from view, leaving a lasting impression with all its viewers (Figs. 6.6 and 6.7).

1P/Halley 1910

A race was on among the astronomical community to be the first to telescopically recover Halley's Comet on its return to the inner Solar System. Its orbit had been fairly well predicted, so astronomers had a good idea of the region of sky towards which they should point their instruments. Astronomers were also aware that this would be a particularly favorable apparition of Halley's Comet, especially for observers in the southern hemisphere. Never before had so much media attention been paid to a comet; people around the world had been to some extent hyped by the magnificent Great Comet that had appeared earlier in 1910.

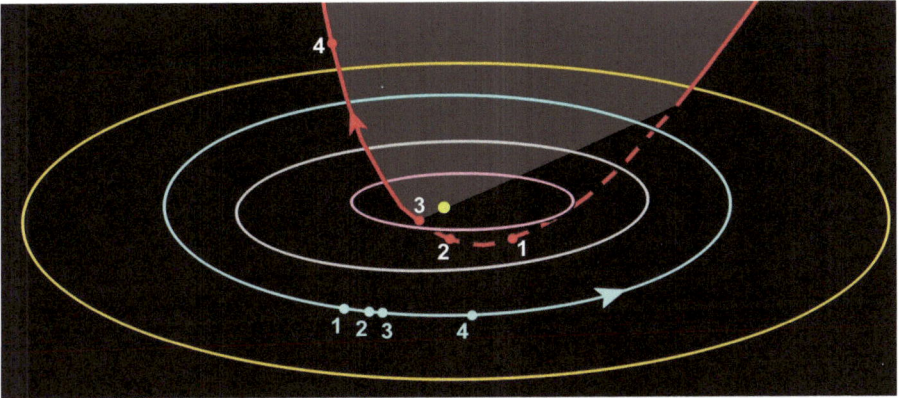

Fig. 6.7 The Great Comet of 1910, showing its path through the inner Solar System (*red line*: above ecliptic; *dashed red line*: below ecliptic) along with the orbit and positions of Earth (*blue*), and the orbits of Mars (*orange*), Venus (*gray*) and Mercury (*pink*). Key: (1) January 12, 1910; (2) January 17; (3) January 18; (4) February 7. (Illustration © by the author)

Halley's Comet was first detected on its inbound course on the morning of September 12, 1909, 7 months prior to perihelion. This prestigious honor went to Max Wolf at Konigstabl, Germany (who was in fierce competition with Edward Barnard in the United States), when he spotted the comet as a 16th magnitude spot in northeastern Orion. Wolf, however, didn't discover it visually at the telescope eyepiece. He found it on a photograph that he had taken, the first Great Comet to have been discovered photographically.

Astronomers followed the course of Halley's Comet as it headed west through the evening skies, gradually brightening and becoming visible through both amateur and professional telescopes in the western skies after sunset. By mid-March 1910 the comet had become lost in the evening twilight, but in early April it had reappeared in the morning skies and soon became visible to the unaided eye. At its lengthiest the tail stretched for 120°, so while it was not as bright as the Great Comet that preceded it, its tail was more than double its length. Perihelion on April 20 saw it pass 0.59 au (88 million kilometers) from the Sun—a comfortable 29° west of the Sun.

Mark Twain's famous lines were to come true. "I came in with Halley's Comet in 1835. It is coming again next year, and I expect to go out with it," said Twain. "It will be the greatest disappointment of my life if I don't go out with Halley's Comet. The Almighty has said, no doubt: Now here are these two unaccountable freaks; they came in together, they must go out together." The great writer died just a day after the comet's perihelion.

On May 18–19 the comet passed directly between Earth and the Sun, our planet actually passing through its tail on the 19th. Prior to this event, which had been predicted long in advance, the famous French astronomer and science popularizer Camille Flammarion is said to have commented that cyanogen gas found spectroscopically in the tail of Halley's Comet would likely 'snuff out' all life on Earth. (Combined in a salt, cyanogens produces cyanide.) Of course, the gases in the comet's tails are so rarefied that absolutely no harm would befall any Earth creature.

Regardless of whether Flammarion actually publicized such an alarming prospect, worldwide consternation ensued. Gas masks sold for inordinately high prices, 'comet pills' were produced to supposedly counter the effects of the comet's noxious emanations and various people and groups—some of whom ought to have known better—prepared for the end of the world. Nothing happened, of course.

Halley's Comet reached its perigee of 0.15 au (22 million kilometers) from Earth on May 20, which coincided with its maximum brilliance of magnitude 0. Under the headlines 'Halley's Comet Speeding Away: Passed between Earth and Sun Wednesday: Caused no convulsions of nature or apparent atmospheric disturbances,' the *San Juan Islander* of May 20 reported:

> Halley's comet, which has been the subject of world-wide discussion for months past and whose coming within the range of human vision after an absence of 75 years had long been awaited with interest by everyone who heard of it and with superstitious dread and awe by great numbers of people, passed between the Earth and the Sun on Wednesday night without causing any known convulsions of nature or apparent atmospheric disturbances. On that day the Earth was enveloped in the comet's tail for a few hours and passed through, or across, about 1 million miles of it. The head of the comet was then distant about 13,000,000 miles from the Earth, as compared with 5,000,000 miles at the time of its last previous appearance in 1835. It is now speeding away into space at tremendous speed, estimated at 1,626 miles per minute, and will not return for 75 years. While it has been visible to the naked eye in the early morning at some points along the Pacific coast for the past week or more, in the eastern sky, it should, astronomers say, be visible in the western sky in the evenings for some days after today. The best view will be obtainable next Monday evening, May 23, when there will be a total eclipse of the Moon, probably between 9 and 10 p.m., thus enabling us to see both the head and tail of this wonder of the heavens in a dark sky for an interval of 15 min. The head of the comet will be brightest today, but the strong moonlight will dim the nebulous light of the tail so that it may not be visible.

Halley's Comet was visible with the unaided eye into July 1910, a total of 80 days (Figs. 6.8, 6.9, and 6.10).

C/1927 X1 (Skjellerup-Maristany)

Largely a southern hemisphere spectacle, this comet's discovery is attributed to two independent amateur astronomers—John Skjellerup in Australia and Edmundo Maristany in Argentina—even though Skjellerup's sighting took place on November 28, 1927, preceding Maristany's observation by 8 days. Like the Great Comet of 1901, the Great Comet of 1927 was noted for its distinct yellow coloration, now known to be caused by emission from sodium atoms.

Fig. 6.8 Halley's Comet photographed from Arequipa, Peru, on April 21, 1910

Becoming visible with the naked eye on November 27, 1927, C/1927 X1 rose to its peak brightness on December 8, a Great Comet shining at first magnitude, and remained easy to spot throughout December. It came closest to Earth on December 12, at 0.75 au (112 million kilometers). Forward scattering of sunlight by the comet meant that it was possible to see the comet with the unaided eye in daylight between December 15 and 18—providing that the Sun, which was just a few degrees away, was safely obscured from view. On December 18 the comet experienced a perihelion of 0.18 au (27 million kilometers) from the Sun. Interestingly, this was the first comet to have been observed in infrared light when Carl Lampland, observing from the Lowell Observatory, took a series of measurements in broad daylight with the 42-in. reflector and a stellar radiometer between December 16–19. At the end of December the comet displayed a tail around 40° long. After fading below naked-eye visibility C/1927 X1 was followed telescopically until March 29, 1928 (Fig. 6.11).

Fig. 6.9 On June 5, 1910, Halley's Comet experienced a magnetic logjam, causing a disconnection event; the comet briefly lost its plasma tail, leaving it far behind the comet's continuing track. (Photograph from the Yerkes Observatory)

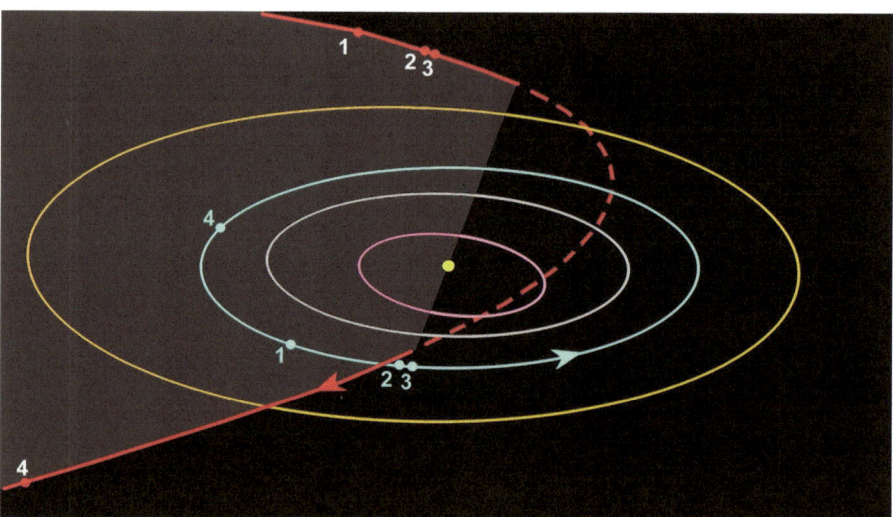

Fig. 6.10 Halley's Comet in 1910, showing its path through the inner Solar System (*red line*: above ecliptic; *dashed red line*: below ecliptic) along with the orbit and positions of Earth (*blue*), and the orbits of Mars (*orange*), Venus (*gray*) and Mercury (*pink*). Key: (1) April 20, 1910; (2) May 19; (3) May 20; (4) February 7. (Illustrated by the author)

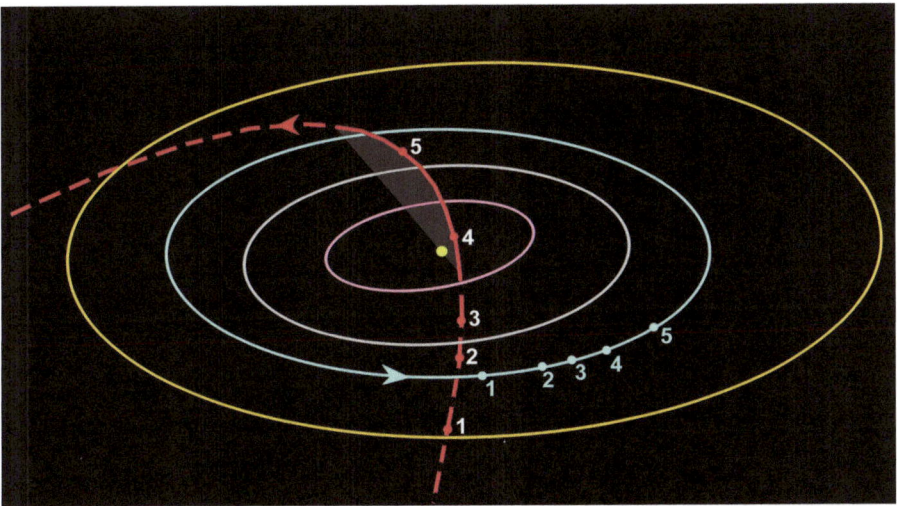

Fig. 6.11 The Great Comet of 1927, showing its path through the inner Solar System (*red line*: above ecliptic; *dashed red line*: below ecliptic) along with the orbit and positions of Earth (*blue*), and the orbits of Mars (*orange*), Venus (*gray*) and Mercury (*pink*). Key: (1) November 28, 1927; (2) December 8; (3) December 12; (4) December 18; (5) December 30. (Illustration © by the author)

C/1965S1 (Ikeya-Seki)

Comet Ikeya-Seki was discovered independently by Japanese amateur astronomers Kaoru Ikeya and Tsutomu Seki (both of whom had made previous comet discoveries) on the morning of September 18, 1965. Soon after the announcement of their discovery it was calculated that the comet would make an extremely close pass by the Sun later that month, and it therefore had the prospect of becoming extremely bright. Indeed, Comet Ikeya-Seki was destined to become the first Great Comet since 1927.

C/1965S1 first became a naked-eye object on October 1, and as it approached the Sun it brightened rapidly. Its perigee of 0.91 au (136 million kilometers) took place on October 17 when it shone at first magnitude. Four days later it reached a perihelion of 0.008 au (just 1.2 million kilometers) from the Sun; it was local noon in Japan, and there were many reports of it being seen at this time (with the Sun suitably obscured), when it reached a heady magnitude of −10. Just half an hour before perihelion, the comet, a member of the Kreutz Sungrazer family, was observed to fragment into three pieces, each of which proceeded in near identical orbits. Re-appearing in the morning sky in late October, the comet displayed a bright, slightly curving tail that extended to up to 25° long. Comet Ikeya Seki was last observed on January 14, 1966 (Figs. 6.12 and 6.13).

Fig. 6.12 Comet Ikeya-Seki. Image from NASA/Stardust website

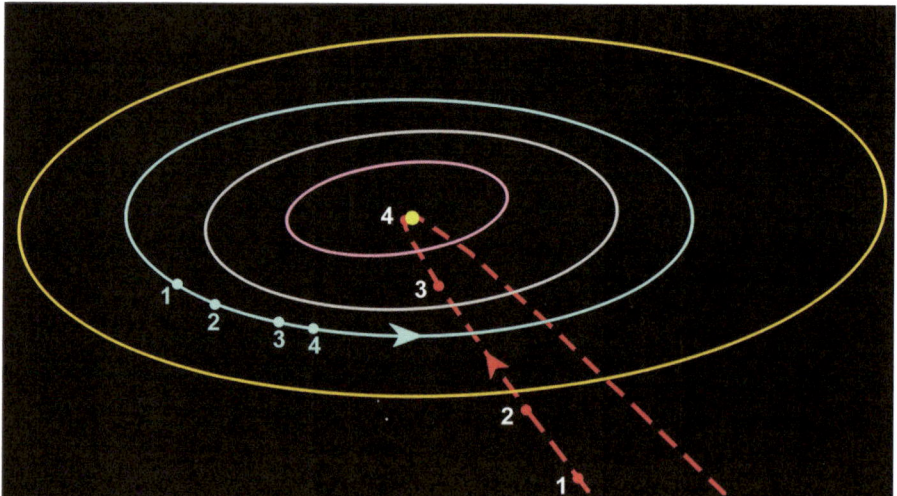

Fig. 6.13 Comet Ikeya-Seki, showing its path through the inner Solar System (*red line*: above ecliptic; *dashed red line*: below ecliptic) along with the orbit and positions of Earth (*blue*), and the orbits of Mars (*orange*), Venus (*gray*) and Mercury (*pink*). Key: (1) September 18, 1965; (2) October 1; (3) October 17; (4) October 21. (Illustration © by the author)

C/1969 Y1 (Bennett)

Comet Bennett was discovered on December 28, 1969, by amateur astronomer John Bennett in Pretoria, South Africa. Bennett had only been at the eyepiece for 15 min that night when the comet—a small, diffuse object glowing at magnitude +8.5—came into view. During the course of January the comet's brightness steadily increased, and by the month's end shone at magnitude 7. Developing a small tail, it became a naked-eye object on February 10, and by the end of the month was shining at magnitude 3.

In mid-March the tail had grown to about 10° long, developed a curve and displayed a fascinating filamentary structure; jets of material emanating from the sunlit side of the nucleus could be clearly discerned in the coma. March 20 saw a perihelion of 0.54 au (81 million kilometers) and 6 days later approached Earth to within 0.69 au (103 million kilometers), shining at magnitude 0. April saw the comet fade from magnitude 1–5, but the tail remained long and curved, with a number of streamers in the immediate vicinity of the coma. Early May saw the comet's 80-day run of naked-eye visibility come to an end; by the end of the month it shone at the 9th magnitude, although the tail could still be telescopically traced to about 2½° in length. Comet Bennett was last observed on February 27, 1971, by which time it was 5.3 au (793 million kilometers) from Earth (Figs. 6.14 and 6.15).

Fig. 6.14 Comet Bennett. Illustration © by the author

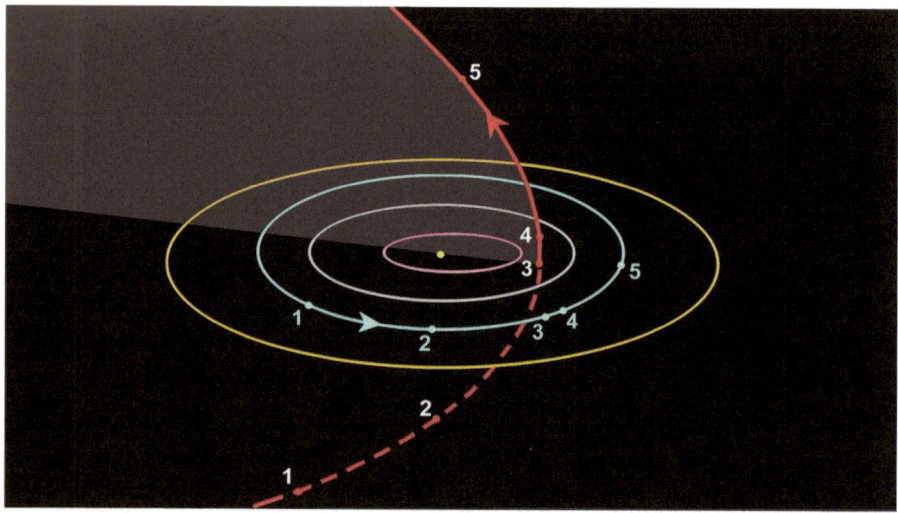

Fig. 6.15 Comet Bennett, showing its path through the inner Solar System (*red line*: above ecliptic; *dashed red line*: below ecliptic) along with the orbit and positions of Earth (*blue*), and the orbits of Mars (*orange*), Venus (*gray*) and Mercury (*pink*). Key: (1) December 28, 1969; (2) February 10; (3) March 20; (4) March 26; (5) 1 May. (Illustration © by the author)

C/1975V1 (West)

Comet West was discovered on a photographic plate taken by the 1-m Schmidt telescope at the European Southern Observatory at La Silla, Chile. Although the photograph had been taken on September 24, 1975, its examiner, Richard West, viewed the image on November 5 and duly found the comet that came to bear his name.

Comet West was largely visible from the southern hemisphere during its apparition. It hit the 8th magnitude in mid-January 1976, when it showed the first signs of developing a tail. The comet became visible without optical aid on February 5, and was to remain a naked-eye object for 55 days. By February 19 it shone at magnitude 1. Its perihelion of 0.20 au (30 million kilometers) from the Sun on February 25 was followed 4 days later by a perigee of 0.79 au (118 million kilometers); at around this time it shone a brilliant magnitude −2.5 and was observable during the daytime.

As it moved away from the Sun, Comet West's tail grew in length. At its longest, around March 7, its dust tail was around 25° long, while its ion tail stretched to about 6°. Close telescopic scrutiny revealed that the nucleus had split into four components. The comet slipped below naked-eye visibility on March 23, and the last view of Comet West was had on September 25, when three of the four nuclei were still visible (Figs. 6.16 and 6.17).

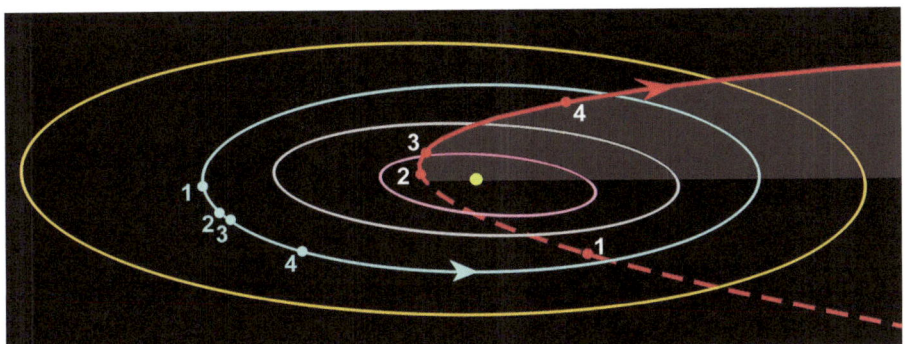

Fig. 6.16 Comet West, showing its path through the inner Solar System (*red line*: above ecliptic; *dashed red line*: below ecliptic) along with the orbit and positions of Earth (*blue*), and the orbits of Mars (*orange*), Venus (*gray*) and Mercury (*pink*). Key: (1) February 5, 1970; (2) February 25; (3) February 29; (4) March 23. (Illustration © by the author)

Fig. 6.17 Comet West in February 1970. NASA image

C/1996 B2 (Hyakutake)

On January 30, 1996, Japanese amateur astronomer Yuji Hyakutake discovered the comet that bears his name; he was using a pair of 25×150 'light bucket' binoculars. Fortunate viewers in the northern hemisphere followed the comet with great enthusiasm.

Hyakutake rapidly reached naked-eye visibility on March 15 as it sped through the constellations of Boötes, Ursa Minor and Cassiopeia during March. Perigee took place on March 25 when the comet was 0.10 au (15 million kilometers) from Earth. Comet Hyakutake became its most brilliant in late March, when it had developed a large coma some half a degree across and a tail that could be traced for 25° from dark sky sites. The comet was easily visible to the naked eye from light-polluted urban areas, and using binoculars the comet appeared to have a distinct greenish hue.

High above the polluted atmosphere the Hubble Space Telescope was swung into action and managed to obtain some exquisite views of the region around the nucleus, showing fine jets of dust and gas, as Hyakutake passed only 9.3 million kilometers from the Earth. The comet was (then) the finest comet to have appeared since Comet West of 1976.

May 1 saw the comet reach a perihelion of 0.23 au (34 million kilometers) from the Sun, after which northern hemisphere observers were only able to view it in morning twilight skies, but those in the southern hemisphere were able to enjoy its journey back into the depths of space. By July the comet had faded from unaided view. Comet Hyakutake was last imaged—shining a dim magnitude 16.8—on October 24 by Gordon Garradd from Australia.

To professional astronomers Comet Hyakutake was noteworthy for its emission of copious quantities of methane and ethane gas—the first time the latter had been detected in a comet. Proportionately the quantity of these substances was measured to be 1,000 times the amount generally thought to have been present in the early solar nebula. This apparent anomaly prompted speculation that Hyakutake had originated in another part of the galaxy, where the mix of raw comet building materials was different than it had been in our own neighborhood.

Adding further to the unusual nature of Hyakutake, the joint ESA/NASA SOHO satellite observatory, from its orbiting post at a gravitationally stable area, called a Lagrange point, some 1.5 million kilometers from Earth, observed tremendous quantities of water vapor pouring off Hyakutake's nucleus. The comet was losing no less than 3 tons of water vapor every second as the icy nucleus sublimated on its approach to the Sun—enough water to meet the demands of a medium-sized industrial town (Figs. 6.18 and 6.19).

C/1995 O1 (Hale-Bopp)

If astronomers thought that they had been treated to an unbeatable once-in-a-lifetime spectacle with Hyakutake, they were in for a surprise when Comet Hale-Bopp began to advertise its celestial presence to all in the months following February 1997.

Fig. 6.18 Author's observation of Comet Hyakutake on April 3, 1996

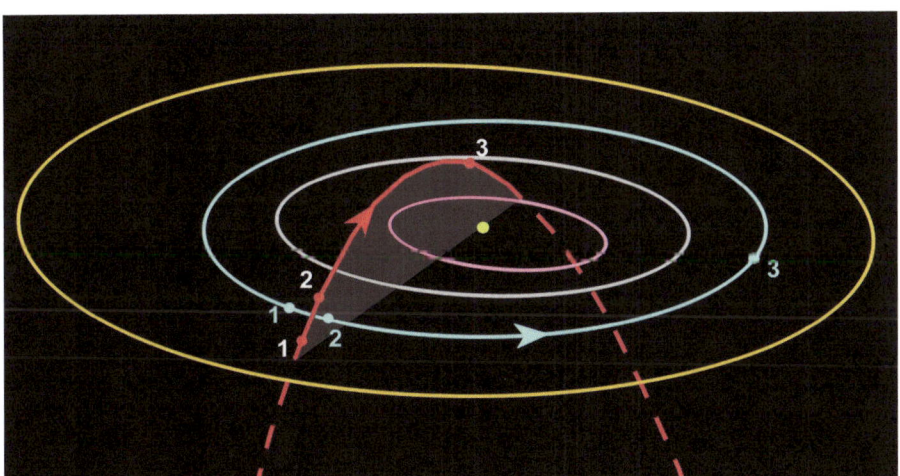

Fig. 6.19 Comet Hyakutake, showing its path through the inner Solar System (*red line*: above ecliptic; *dashed red line*: below ecliptic) along with the orbit and positions of Earth (*blue*), and the orbits of Mars (*orange*), Venus (*gray*) and Mercury (*pink*). Key: (1) March 15, 1996; (2) March 25; (3) May 1; (4) July 1. (Illustration © by the author)

Comet Hale-Bopp was discovered independently on July 23, 1995, by American astronomers Alan Hale (a professional) and Thomas Bopp (an amateur) when it was a faint magnitude 10.5 patch of light in the middle of Sagittarius. It was soon found to be a very distant incoming object, well beyond the orbit of Jupiter at discovery. Further investigations showed that the comet had actually been imaged by the UK Schmidt Telescope at Siding Spring in Australia as far back as April 1993, when it was a mere 18th magnitude.

Being so bright and yet so far out at the time of its discovery, Hale-Bopp was allocated a brightness curve prediction of stunning proportions. Astronomers began to get excited at talk of a magnitude −1.7 coma (brighter than Sirius, the brightest star in the sky) at the time of its closest approach to Earth, a respectful 1.32 au (197 million kilometers) on March 22; perihelion took place on April 1, at 0.91 au (136 million kilometers) from the Sun. Astronomers were cautious in accepting even the most pessimistic predictions because of past experiences with other comets that had failed to live up to expectations.

Thankfully, Hale-Bopp was different. By late 1996 the visitor had brightened enough to be seen through binoculars in the western evening skies after sunset. The comet was lost to sight for a short while as it became immersed in the glow of sunset, but by early 1997 it had cleared the Sun and for the first time had become visible to the naked eye, tail and all, as it passed through the constellation of Cygnus.

Astronomers in the northern hemisphere could not believe their eyes as the true claimant to the title of 'Comet of the Century' brightened and established itself as a prominent circumpolar object. This time, the various news media were not raising false hopes when they encouraged people to go outside and spot the comet high in the western evening skies. From urban sites the comet easily shone through the perpetual orange blanket of the streetlights' glow, and the brightest part of the tail, around a degree or two long, was easy to see. From darker sites the tail could be discerned as having two components—a curved yellowish dust tail approaching 10° in length and a longer straight blue gas tail.

Telescopic views of Hale-Bopp's coma showed a bright star-like nucleus surrounded by several distinct arcs of light—shells of dust and gas being swept back by the solar wind. The Hubble Space Telescope secured some fantastic close-up images of the jets arising from the 60-km-diameter rotating nucleus.

Comet Hale-Bopp was visible to the naked eye for a record time, twice as long as the previous record holder, the Great Comet of 1811. The comet had remained visible without aid for 569 days, or about 18½ months (Figs. 6.20, 6.21, and 6.22).

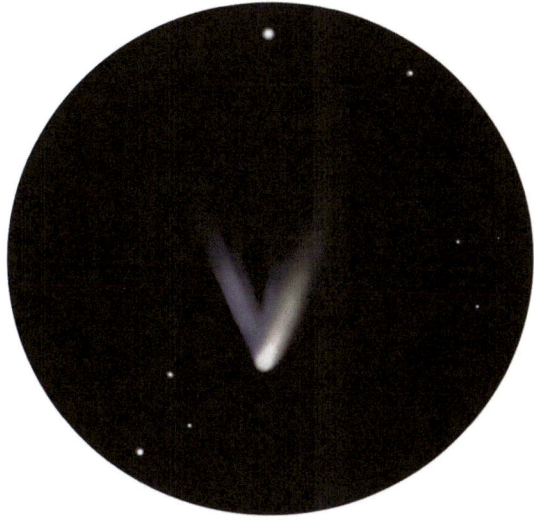

Fig. 6.20 Author's observation of Comet Hale-Bopp, made on February 25, 1997

Fig. 6.21 Comet Hale-Bopp, imaged on March 14, 1997. ESO/E. Slawik

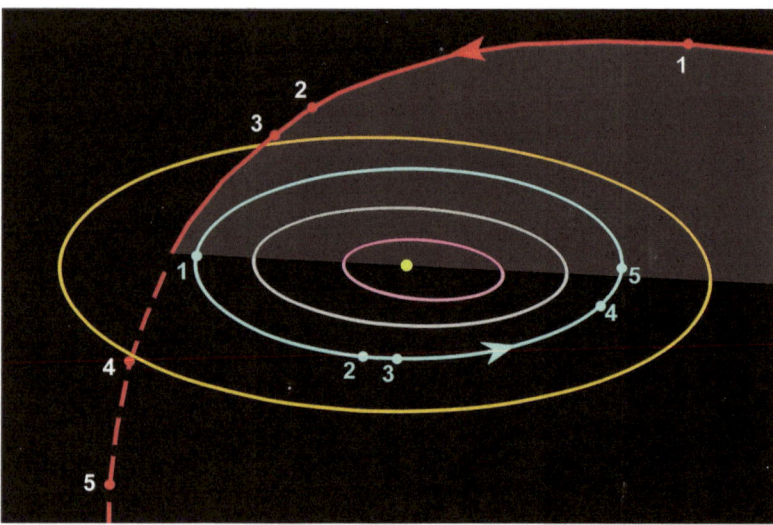

Fig. 6.22 Comet Hale-Bopp, showing its path through the inner Solar System (*red line*: above ecliptic; *dashed red line*: below ecliptic) along with the orbit and positions of Earth (*blue*), and the orbits of Mars (*orange*), Venus (*gray*) and Mercury (*pink*). Key: (1) March 22, 1997; (2) April 1. (Illustration © by the author)

Chapter 7

Locating, Tracking, and Recording Comet Sightings

There are few celestial sights as awesome as a bright comet, complete with tail(s), suspended in the night sky—a luminous heavenly wraith, a glowing apparition clearly in flight but seemingly frozen in time, a phantom of the Solar System, the likes of which understandably in earlier times struck more ignorant generations of viewers with feelings of foreboding, awe and terror. Nowadays we still feel the awe. Regardless of our knowledge, humans (even scientists!) are not immune to nature's splendors, but our adrenal glands are stimulated by the sheer pleasure of the view, not by superstitious fears.

Locating Comets

Unless a comet is particularly bright and prominent, easily visible with the unaided eye in the night sky, you're going to need to know where to point your binoculars or telescope in order to see it. That will entail either consulting a star chart with a plot of the comet's predicted path (points along the path being marked with a date) or consulting an ephemeris, a dated list of celestial positions given in right ascension (RA) and declination (Dec) (Fig. 7.1).

Ephemerides

Ephemerides give the predicted position of a comet's nucleus in terms of its RA and Dec in set intervals of time (usually for 00 h UT on each given date). Most ephemerides

P. Grego, *Blazing a Ghostly Trail: ISON and Great Comets of the Past and Future*, The Patrick Moore Practical Astronomy Series, DOI 10.1007/978-3-319-01775-4_7, © Springer International Publishing Switzerland 2014

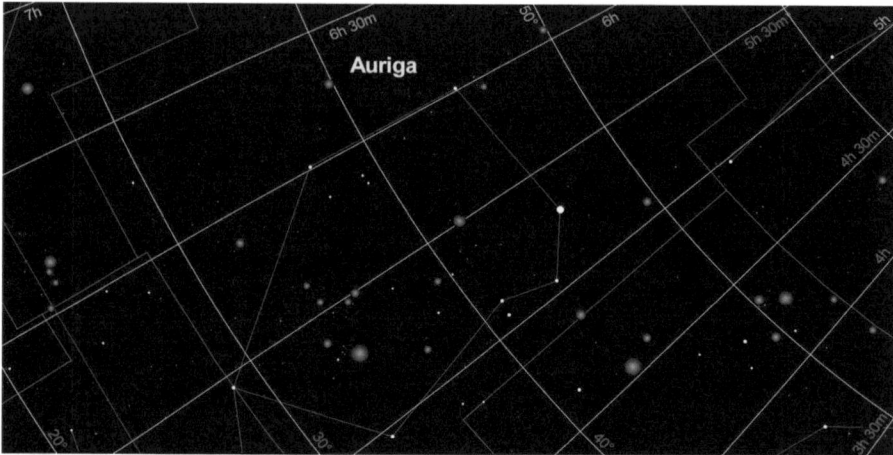

Fig. 7.1 Where is the comet? One of these 'faint fuzzies' in the constellation of Auriga is 12th magnitude comet C/2012S1 (ISON) at 21 h UT on April 4, 2014. The comet is located at around RA 5 h 30 m, Dec 41° 35′. (Illustration © by the author)

also give some indication of how bright the comet is expected to be on each date, based on early observations and typical cometary light curves. As more positional measurements of a newly discovered comet are made, its parameters become more refined and reliable. It is common to see the RA and Dec figures given in an initial ephemeris differing slightly from those in an updated ephemeris, along with predictions of a more realistic light curve (Fig. 7.2).

Charts and/or ephemerides of currently visible, soon-to-be-visible or newly discovered comets are featured in a variety of astronomical publications, such as *Popular Astronomy* (Society for Popular Astronomy, UK), the annual *Handbook* of the British Astronomical Association (UK), as well as newsstand magazines such as *Sky % Telescope* and *Astronomy* (U.S.), *Astronomy Now* and *Sky at Night* (UK) and *Australian Sky & Telescope*. In addition, there's no shortage of websites that feature finder charts and ephemerides for comets; NASA/JPL's Solar System Dynamics Horizons web interface at http://ssd.jpl.nasa.gov/horizons.cgi is a highly recommended resource (Fig. 7.3).

Celestial Atlases

A good star atlas showing all the naked-eye stars and made to a generous scale is indispensable to the comet observer, even in these days of computerized planetarium programs. *Norton's Star Atlas 2000.0* is a good all-around resource, with its sixteen star charts giving complete sky coverage and depicting 8,800 stars down to

INDIVIDUAL COMETS

2P/Encke

2013	RA	(J2000.0)	Dec.		Mag.	Δ	r	Meridian Transit	Elong. (°)		Moon
0h TT	h	m	°	'		AU	AU	hh:mm	Sun	Moon	Ph.%
Sep.											
4/5	4	15.5	+34	44	11.5	1.07	1.52	05:19	94	91	0
14/15	4	55.9	+38	17	10.8	0.87	1.39	05:20	95	139	76
24/25	5	57.3	+41	57	10.0	0.69	1.24	05:42	92	29	70
Oct.											
4/5	7	36.1	+43	27	9.1	0.55	1.09	06:42	84	85	0
14/15	9	48.2	+36	49	8.4	0.48	0.92	08:14	67	149	82
24/25	11	40.5	+20	55	7.9	0.51	0.75	09:27	47	69	69
Nov.											
3/4	12	54.2	+4	54	7.7	0.65	0.56	10:01	32	39	0
13/14	13	50.3	-7	41	7.3	0.88	0.39	10:18	23	160	87
23/24	14	53.9	-17	57	7.6	1.17	0.34	10:42	15	94	67
Dec.											
3/4	16	6.0	-24	36	8.8	1.42	0.47	11:15	8	23	1
13/14	17	9.9	-27	34	10.0	1.62	0.65	11:39	6	146	90
23/24	18	2.8	-28	21	11.0	1.80	0.83	11:53	5	105	66

Fig. 7.2 Ephemeris for 2P/Encke in 2013, as given in the annual *Handbook* of the British Astronomical Association. (Courtesy of the British Astronomical Association)

magnitude 6.5 (just fainter than the eye can see). It also shows 600 of the brightest deep-sky objects, including the entire Messier list and many NGC objects. In terms of scale, *Norton's* averages 3.2 mm per degree (3.1° per cm); this doesn't sound much, but it's just about right for a star atlas that can be comfortably used in the field as well as to consult indoors (Fig. 7.4).

A more detailed set of 26 celestial charts (measuring 21 × 16 in.) is contained in *Sky Atlas 2000.0*. This contains 81,000 stars down to magnitude 8.5, shows 2,700 deep-sky objects and has a scale averaging 8.2 mm per degree (1.2° per cm). Going one step further, *Uranometria 2000.0* shows around 280,000 stars down to magnitude 9.8 and more than 30,000 deep-sky objects; with 220 charts (measuring 18 × 12 in.) that average 18.5 mm per degree (0.6° per cm), this is the most detailed general star atlas in print (Fig. 7.5).

Based on ephemeris positions, a comet's predicted position can be plotted in pencil upon a suitable star chart (or a photocopy of one if you prefer not to deface your original). If plotting the comet's predicted path, make sure that small tick marks at intervals are labeled with date and time. Once a path is plotted, the observer is free to interpolate (estimate) a comet's position between dated tick marks. My own vintage and well-used copy of *Norton's Star Atlas*, for example, has at least one cometary path marked on each page, including such notable visitors as Iras-Araki-Alcock of 1983, Halley's Comet, Brorsen-Metcalf of 1989, Swift-Tuttle of 1992. In fact, I've penciled in virtually every reasonably bright comet visible from my location from the 1980s into the mid-1990s. That particular practice was abandoned when I was introduced to the wonderful capabilities of

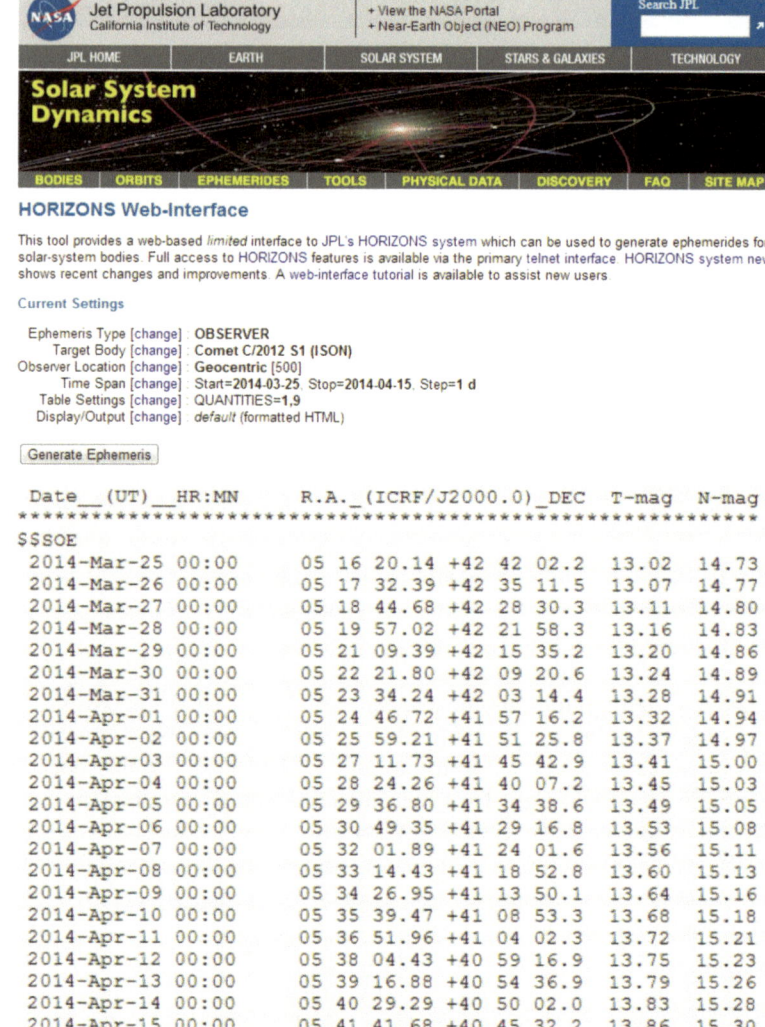

Fig. 7.3 Web-based search on the Horizons interface for the position and predicated magnitude of C/2012S1 (ISON) between March 25 and April 15, 2014, and the ephemeris generated by the query. (NASA/JPL)

astronomical programs for the personal computer, but I still consult *Norton's* and *Sky Atlas 2000.0* when in the field. Although technology is wonderful, there's something fulfilling about handling a printed chart while at the eyepiece, poring over it with a red torch. Moreover, an atlas won't break if you accidentally drop it in the dark!

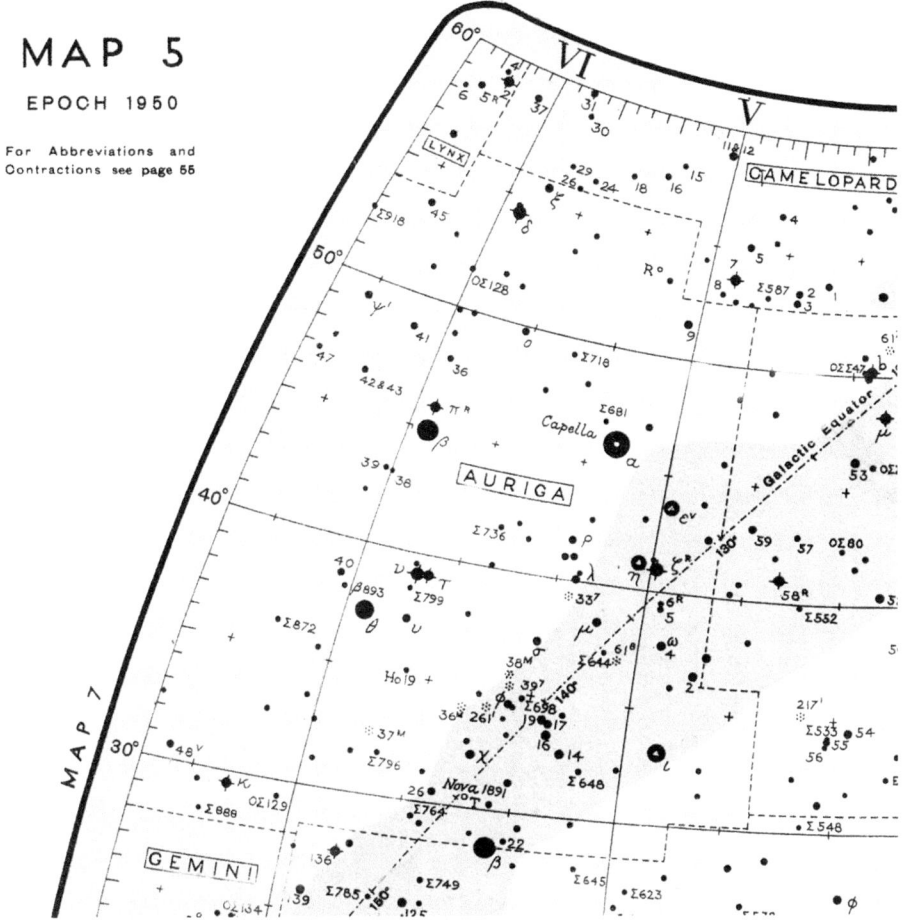

Fig. 7.4 Close-up of the Auriga region, as depicted on Chart 5 of the 1957 edition of *Norton's Star Atlas* (Epoch 1950.0). (From the author's collection)

Planetarium Programs

Many of the better astronomical programs for personal computers, tablets and phones are capable of displaying the positions of known periodic comets with a high degree of accuracy, in addition to giving information about the comet's predicted magnitude and other data.

The website http://cfa-www.harvard.edu/iau/Ephemerides/Comets/Software Comets.html gives the orbital elements of observable comets in formats suitable for loading into a number of popular planetarium-type software packages.

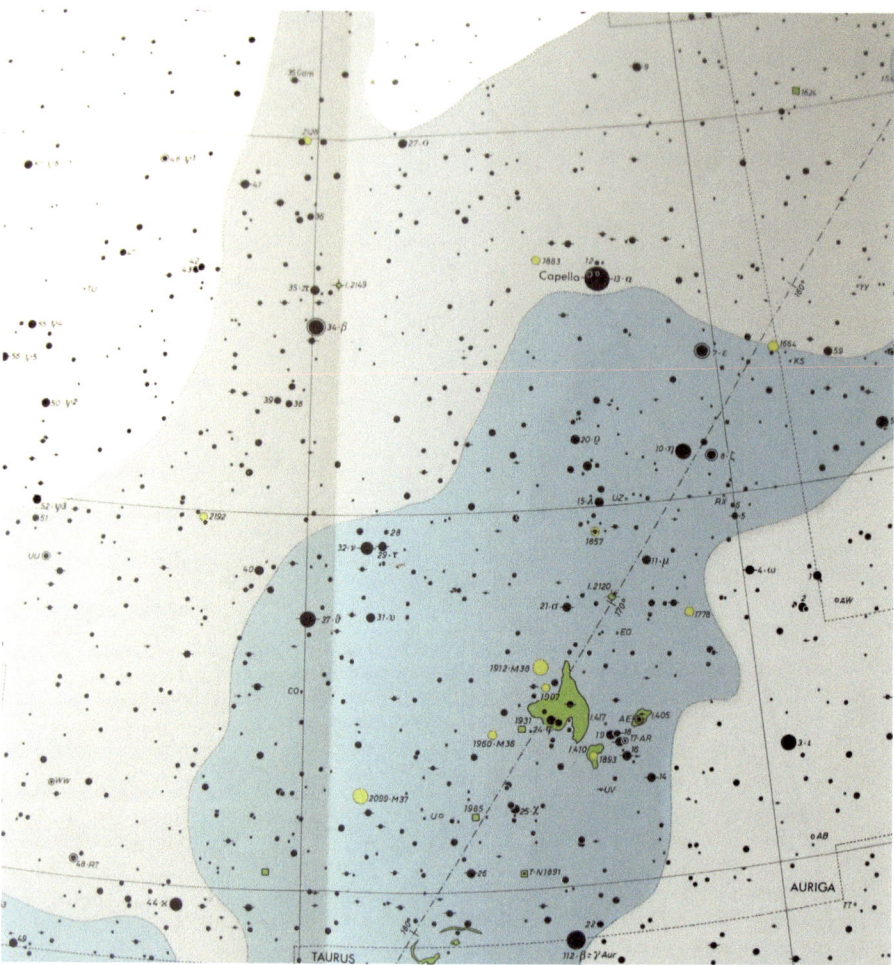

Fig. 7.5 Close-up of the Auriga region, as depicted on Chart 5 of *Sky Atlas 2000* (Epoch 2000.0). (From the author's collection)

To ensure that the latest ephemerides are available, *Starry Night*, in common with many other good computer programs, has a live update facility where the latest data on the positions of periodic comets and newly discovered comets can be downloaded automatically. It is also possible to download data on the orbital parameters of newly discovered comets and manually input the figures into a variety of planetarium programs (Fig. 7.6).

Computer programs are a great observational and research tool, but they are only as good as the data available to them. Although they will show a comet according to its predicted brightness, displaying a nice graphic image of the comet and the general direction of its tail, they are not sophisticated enough to be able to predict

This file kindly prepared by the IAU Minor Planet Center & Central Bureau for Astronomical Telegrams.

Num	Name	Mag.	Diam	e	q	Node	w	i	Tp	Epoch	k	
	Spacewatch	16.0	0.0	0.485760	2.125103	161.2218	22.1142	14.4665	2456389.6853	2456501.5	5.0	283P
	McNaught	13.0	0.0	0.376996	2.289123	144.3013	202.8708	11.8622	2456903.0031	2456501.5	10.0	284P
	LINEAR	15.0	0.0	0.623567	1.690752	186.3886	177.4674	24.6016	2456472.3560	2456501.5	10.0	285P
	Christensen	14.0	0.0	0.423812	2.375976	283.9529	24.8737	17.0214	2456663.6537	2456501.5	10.0	286P
	Blanpain	10.0	0.0	0.684685	0.961106	68.9460	9.8241	5.9004	2456897.6852	2456501.5	10.0	289P
	Jager	10.5	0.0	0.648370	2.155791	303.4317	180.7659	19.0529	2456729.0989	2456501.5	10.0	290P
	NEAT	13.0	0.0	0.430489	2.590960	241.0413	176.0667	5.9569	2456641.9383	2456501.5	10.0	291P

Fig. 7.6 Data for a number of comets for use in *Starry Night's* orbit editor. Don't be put off by all the numbers. Data for new comets can easily be added manually, as long as you take care to put the correct data in the *right boxes*. Many programs use automatic updating to save you the trouble! (Courtesy of the Minor Planet Center)

Fig. 7.7 Comparison between the depiction of Comet Holmes by Starry Night Pro™ and the author's actual binocular observation of the comet on October 28, 2007, at 22:30 UT. (Image © by the author)

the actual behavior of a comet. Apart from the general direction of the comet's tail, the graphics that they display fail to show what the object will actually look like through binoculars or the telescope eyepiece (Fig. 7.7).

Research into previous or future apparitions of comets is another matter entirely, since most commercial planetarium programs assume current orbital parameters and fail to take into account orbital changes caused by the perturbations of planets. For example, *Starry Night Pro Plus* (Ver. 6.4.3) gives the position of Halley's Comet at its perihelion on April 20, 1910, as being in Hydra (calculated using the comet's current parameters), some 155° away from where the comet actually was at the time—on the other side of the sky in Pisces. The program does, however, include separate parameters for the comet's 1910 apparition, but this isn't the case for other periodic comets, so historical research must be conducted with care.

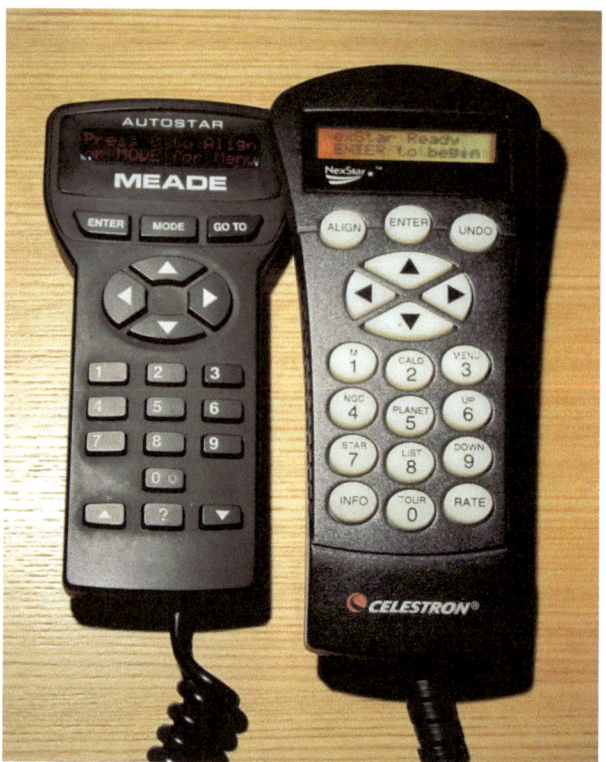

Fig. 7.8 Comparison between the Meade Autostar and the Celestron NexStar handsets. The Autostar gives red LED display, while the NexStar is a backlit LCD display. (Image © by the author)

Computerized Telescopes

Meade and Celestron go-to telescopes are provided with electronic handsets that have a built-in library of ephemerides for any number of periodic comets; the library can be updated via the Internet to contain the most up-to-date data, including new comets. At the touch of a few buttons it is possible to have your telescope slew around to, say, the latest position of Halley's Comet (though it's now so dim and distant that it will certainly be invisible) and to have a list of relevant data about the comet scroll across the handset display. It is also possible to input a specific RA and Dec from an ephemeris and instruct the telescope to slew directly to that position. If the comet is not visible, it is either too faint to see or it is not quite within the field of view. In the latter case, the computerized telescope can be instructed to perform a slow spiral search around the position it is currently pointing towards. All of these facilities are very handy (Fig. 7.8).

Manual Searches

If you're searching for a comet using binoculars or a telescope without a computerized go-to feature, it's best to first of all make sure that the comet's predicted brightness is well within the magnitude grasp of your instrument (see the chapter on Kit for Comet Observing); otherwise there's a good chance that your hunt will prove futile.

Although the predicted brightness of periodic comets is generally reliable, the brightness of newly discovered comets on their first known foray into the inner Solar System can vary considerably from predictions, either failing to meet predictions or in some cases exceeding them. Indeed, some comets—both periodic and newly discovered—have been observed to flare for a period of time, owing to a fresh burst of activity from newly exposed volatiles, either through the direct impact of a sizable object (say a large meteoroid) on the nucleus, fresh fissuring in the nucleus (due to a build-up of gas beneath the surface) or when the nucleus itself breaks apart (as sometimes happens when a Sun-grazing comet is near perihelion). This can increase a comet's brightness by several orders of magnitude for a time. Such a dramatic event—actually, the largest cometary outburst ever recorded—took place in late October 2007 when Comet Holmes (C/17P, discovered in 1892) brightened by a factor of half a million over a period of less than 2 days, raising it from a very faint magnitude 17 object, visible only through large telescopes, to an easily visible naked-eye spectacle of magnitude 2.8.

If the searched-for comet is faint and in a particularly barren part of the sky, you will need to 'star hop' to find the right field of view, rather than making random sweeps of the general area to locate the comet. That's one of the advantages of having a large optical finderscope whose cross-hairs are well-aligned. Some finderscopes deliver a 'telescopic' (upside down) view, which can be a little confusing to use when trying to use a regular star atlas. Right-angle finders are available for those who prefer a 'right way up' view. An indispensable aid to star-hopping is to have some pre-prepared circles drawn on transparent plastic, or a circle formed out of thin wire, whose diameter is equivalent to the apparent field of view of your binoculars or different types of eyepiece for any given atlas or finder chart. Begin with a bright star, asterism or pattern of stars in the comet's vicinity and then use these circles to hop from star to star, taking care to ensure that the orientation of the chart and the direction in which your instrument moves are congruent. Once you have found your comet you can begin to make a drawing on the final chart (Fig. 7.9).

Preparing Your Observing Template

This is the chart upon which you're going to sketch the comet. Many observers prefer to use a blank sheet of paper with a suitably sized circle representing the field of view; the star field surrounding the comet is added manually, based on eyesight alone. Sometimes there are so many stars in the field (often very faint ones) that it's practical only to depict the brighter ones. The original observational drawing can,

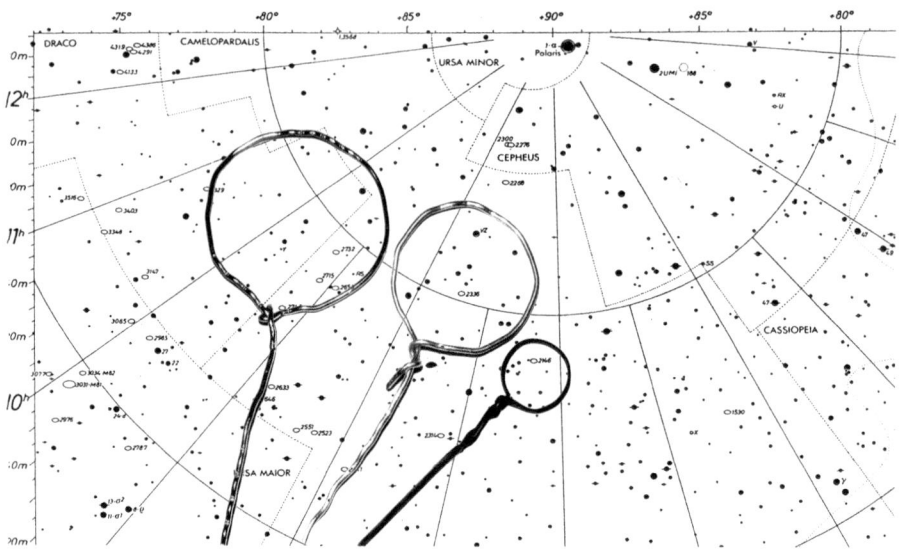

Fig. 7.9 Useful hand-made circles corresponding to 6, 4 and 2° prepared for quickly gauging relative fields of view through 10×50, 15×70 and 25×100 binoculars respectively when using Sky Atlas 2000. (Image © by the author)

of course, be corrected after the observing session so that the star positions are represented accurately.

It is also possible to make your observational drawing of the comet directly onto the finder chart to save time plotting the surrounding star field. Note, however, that a small scale chart such as those in *Norton's* will be difficult to use as an observing template. Even if you scan the chart, enlarge it and print it out to a reasonable size, the stars will appear as large black dots, and all the fainter stars visible through binoculars or the telescope eyepiece won't be shown. If you're preparing a template based on a printed atlas, select an area centered on the comet's position that's larger than the field of view delivered by the instrument/eyepiece that you will be using, and lightly draw upon it a circle corresponding to the field of view.

All of this, however, is best done using a planetarium program in conjunction with a graphics program, rather than manually. The advantages of a computer-generated template include the ability to change the range of magnitudes represented on the chart to suit your instrument, being able to display stars as tolerably small dots regardless of the magnification (some planetarium programs even offer the capability of changing the range of dot sizes representing stellar magnitudes), and giving the 'right way up' orientation for your particular instrument (Fig. 7.10).

Fig. 7.10 A comet finder chart produced in Starry Night Pro™, based on the predicted path of C/2009 P1 (Garradd) and centered on its predicted position on the morning of March 26, 2012. The 4.4° 15×70 field of view with which the visual observation would be made separately is marked for scale. (Image © by the author)

Visual Detail Within the Comet

Bright comets can be enjoyed with the unaided eye. However, light pollution is a great handicap when attempting to discern any of the subtle detail within the coma or tail(s). For example, from the suburbs of a major city, the bright magnitude 0 head of Comet Hyakutake of 1996 could easily be seen without optical aid, but little else was discernable. From a dark countryside location, the comet was splendid with the unaided eye, with a narrow sky-spanning tail some 70° long.

Binoculars, with their wide field of view, are ideal for general comet observing—even then, they may not have a field of view wide enough to take in all of a bright comet's tail. Binoculars are also better at delivering color to the eyes than a telescope eyepiece (Fig. 7.11).

Telescopes are capable of being used at fairly high magnifications to view detail in and around the nuclear regions of a bright comet; the more magnification that is

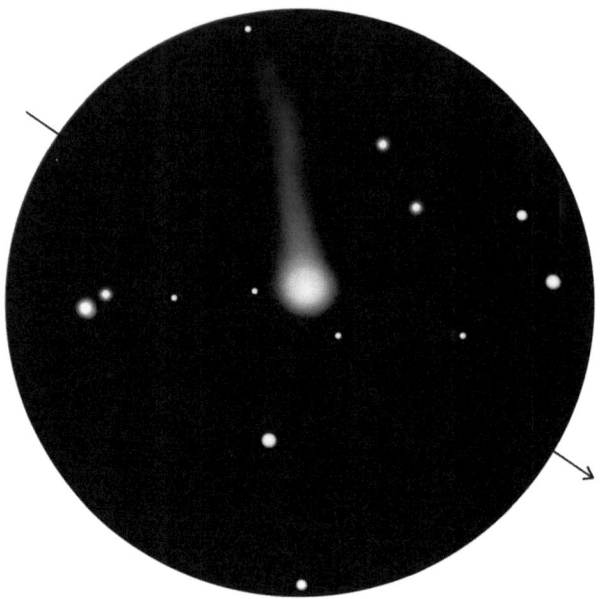

26 October 2006   21h UT
15x70 binoculars   4.4° field
Seeing: All-III   Transp: 4, slight haze
Peter Grego (Rednal, UK)

PDA sketch

Fig. 7.11 Completed observational drawing (using a PDA) showing C/2006M4 (SWAN) at 21 h UT on the evening of October 26, 2006, using 15×70 binoculars (4.4° field of view). (Illustration © by the author)

used, the less the amount of faint detail is visible, but structure within the brightest parts of the nuclear region becomes easier to discern. It's a good practice to use a variety of eyepieces giving a range of magnification—a wide-angle, low power eyepiece for the general appearance of the coma (and tail or part of the tail, if visible), a medium-power eyepiece and a high power one for detail in and around the false nucleus.

Recording Comets

It is tempting to regard pencil drawings as a quaint and somewhat outmoded means of recording the appearance of any celestial object. In the case of comet observing, however, a drawing of what can be seen with the eye through the telescope eyepiece

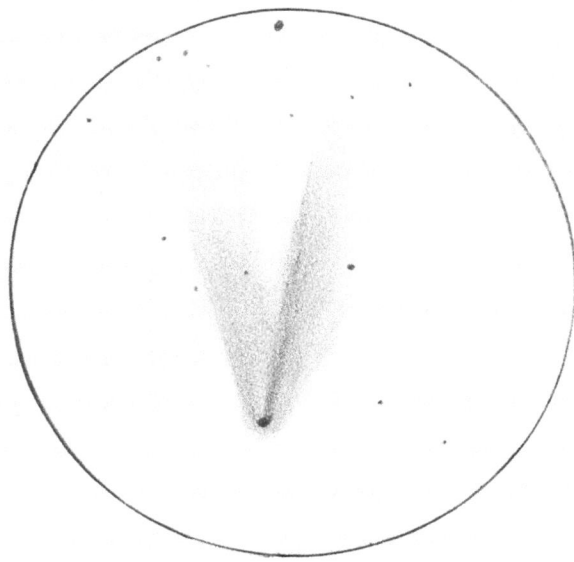

Fig. 7.12 Observational pencil drawing made at the eyepiece showing C/1995 O1 (Hale-Bopp) made at 05 h UT on the morning of March 6, 1997, using 7×50 binoculars. (Illustration © by the author)

can deliver more data than a CCD image. The eye has a very wide dynamic range, capable of taking in subtleties of tone as well as intricate detail. CCD cameras are not perfect; while in a single time exposure they may be able to capture subtle detail and variations in brightness in the fainter tail regions, brighter areas near the nucleus will tend to be burned out. Therefore CCD work can require a lot of processing and image enhancement on the computer, even the combination of separate exposures to bring out a suitable overall view.

The drawing itself is best made on a blank of smooth white paper using a soft pencil. First, mark the position of the bright nuclear condensation; if the whole comet can be comfortably seen within the field of view, place the comet at the center. If the comet is bright and extends out of the field, place the comet's nucleus in a position where the most overall detail can be seen. Next, draw in the brightest stars within the field of view as accurately as possible. If it's possible to identify any of them, note their name. Once the main stars have been sketched, draw in the fainter stars. Returning to the comet, sketch the approximate size and shape of the nuclear brightening. Using layers of lightly applied pencil, draw in the coma and any detail visible within the tail(s). A blending stump or the tip of your index finger serves nicely as a smudger, merging where required, and a sharpened eraser can be carefully used to denote any clear-cut regions of darkness around or within the coma or tail. An alternative to making a 'negative' pencil drawing is to use pastels on dark paper; most prefer the convenience of ordinary pencil (Figs. 7.12 and 7.13).

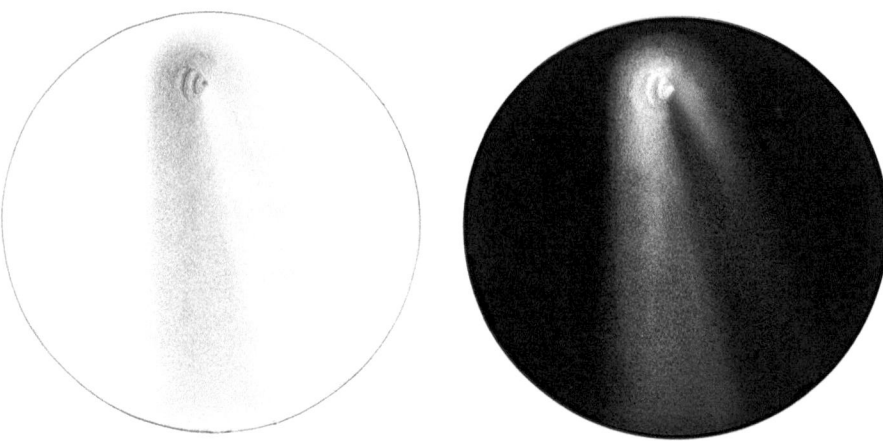

Fig. 7.13 Observational pencil drawing made at the eyepiece of C/1995 O1 (Hale-Bopp) made at 05 h UT on the morning of March 20, 1997, using a 6-in. Newtonian at a magnification of 50. The same image has been scanned and made into a negative at right. Much detail can be seen in the coma. (Illustration © by the author)

Tracking Comets

Tracking a comet visually and plotting its motion against the stellar background may appear to be a very simple project, and many readers might wonder why anyone should bother to do it. After all, it's possible to accurately plot the path of a comet over any period of time using any one of dozens of planetarium programs. But that would be missing the point of amateur astronomy, failing to grasp the reason why tens of thousands of people enjoy observing the night skies. And although it's not a highly technical project, its simplicity is deceptive. Certain guidelines need to be observed in order to achieve success.

At the most basic level the observer could use a blank sheet of white paper (upon which is drawn a circle of, say, 100-mm diameter), center the comet and mark in the surrounding stars as accurately as possible without reference to a star map, producing a separate drawing on each observing session. This is the author's approach when making a one-off low-power observation of a comet; it allows you the challenge of being able to represent what you can see through the eyepiece, including plotting the main stars and representing their relative brightness.

Since our aim is to produce a drawing showing the comet's motion over a period of several days or weeks, incorporating a number of individual observations, the inaccuracies introduced by this method might make for a less than smooth-looking track and a possibly somewhat confusing star field when the raw observations are later combined when producing a finished observation without any form of overall correction (Fig. 7.14).

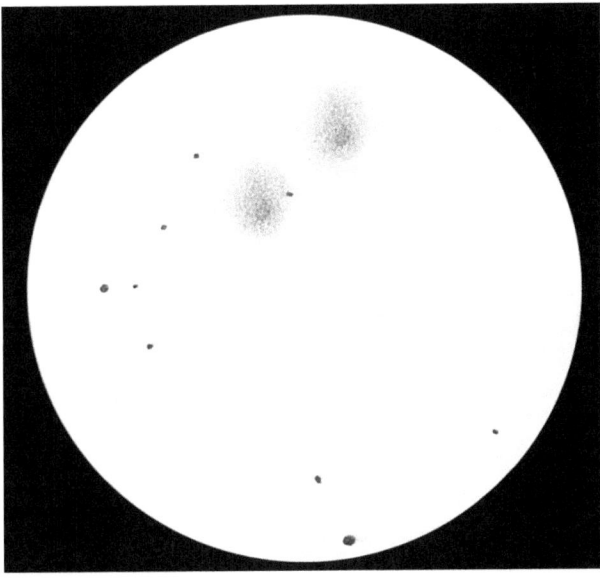

Fig. 7.14 Observational pencil drawing made at the eyepiece showing C/1983H1 (Iras-Araki-Alcock) showing its motion among the stars on May 9, 1983, between 22:30 and 23:00 UT. (Illustration © by the author)

Observational Example: Sequential Comet Observation, No Finder Chart Used

Date: May 9, 1983
Time: 22:30-23:00 UT
Object: C/1983H1 (Iras-Araki-Alcock)
Instrument: 60 mm achromat (undriven), 64× (actual field of view 50 arcmin)
Sketch made with: Pencil (2B) negative sketches

Notes: The comet was easily visible with the unaided eye as an extensive (at least a half-degree wide) glowing patch near Gamma Ursa Minoris, so no finder chart was necessary. Observational drawings were made with unaided eye, 12×50 binoculars, 6×30 finderscope and a 60 mm achromat at 64× and 100×. The comet was moving rapidly across the sky (by about half a degree per hour at this stage), so a sequential observation was made between 22:30 and 23:00 UT using a 13 mm Plossl, giving an apparent field of 50° and an actual field of 50 arcmin. The comet was drawn to the east of the center of the field of view (field centered at around RA 15 h 48 m, Dec 73° 37 min). Estimated magnitude 2.5. At this magnification the size of the coma appeared to be around 20 arcmin wide, ill-defined but with bright single, well-defined nucleus to one side of coma (towards the south); PA of around 350°. No chart was used to plot the stars—the observational drawing is based solely upon at-the-eyepiece observation.

Observational Example: Sequential Comet Observation Using Finder Chart/Observing Template

Date: March 30 and 31, 2013
Time: 21:30 UT (both observations)
Object: C/2011L4 (PanSTARRS)
Instrument: 25 × 100 (actual field of view 2.5°)
Sketch made with: Pencil (2B) negative sketch, enhanced and inverted in *PhotoPaint™*

Notes: The *Starry Night Pro™* program was used to produce a white background star map showing a magnified view of the area in which the comet was located (around 10 × 10°) but without showing the comet itself. The stars shown on the observing blank went down to magnitude +10.0, which is easily reached by the 25 × 100 s.

Observational drawings of the comet and its tails were made on two consecutive nights at the same UT time. The comet was moving at quite a clip at this time (moving 1.3° north between the two observations), but both observations were able to be incorporated into a single image showing both fields of view overlapping.

The observation was scanned in Corel *PhotoPaint™*, overlapping the two fields. To produce the finished drawing, the contrast level on the combined image was lowered and the stars were re-dotted in black using the paintbrush tool, making sure to dot the stars at the same size. The image was then inverted, producing white stars on a black background. Finally, the brightest stars in the field, along with the comet, were Gaussian blurred to reinforce their brilliance and then re-dotted at their centers with a white brush. A slight orange hue was added to the comet's tail (Fig. 7.15).

Cybersketching

Several years ago it became clear that the PDA (Personal Digital Assistant) and tablet PC—both hand-held computers with touch screens and stylus input—had the potential to make observational sketching far easier. Of course, there are downsides. Electronics tend to get sluggish and then refuse to work at very low temperatures, and unless you have a rugged computer, mishaps in the field may cost a great deal more than would a dropped sketchpad. But the tremendous versatility afforded by PDAs and tablet PCs far outweigh their disadvantages, and the experience is most enjoyable.

Fig. 7.15 Observational sequence, made on PDA, showing the motion of C/2011L4 (PanSTARRS) among the stars on March 30 and 31, 2013. (Illustration © by the author)

Basic Freehand Cybersketches

If you're completely new to cybersketching, unsure about using layers in a drawing, and prefer to make a cybersketch with just the basic tools in a graphics program, here are a few simple tips for producing a freehand field cybersketch of a comet on PDA or tablet PC:

- Field sketches of comets are easier to draw in negative, i.e., black stars on a white background, brighter regions of the comet appearing dark gray (final tonal adjustment can be made indoors). To conserve dark adaptation the brightness of the device needs to be set low; alternatively, a red background can be used instead of a white one, as it's a simple matter to change the background color when producing the finished sketch indoors at the computer.
- If possible, draw a circle to represent the edge of your binocular/eyepiece field of view and keep your sketch within its boundaries. This helps maintain your orientation while sketching.
- If the comet is an extended nebulous object, draw it lightly with a light gray shade and then blur the image, or smudge it around the edges. Additional brushstrokes of varying gray tones can be added on top, each tone blurred repeatedly to achieve the right look.

Fig. 7.16 Example of an observational drawing (*left*) made on a hand-held computer (HP iPAQ hx4700) showing C/2007N3 (Lulin) on February 21, 2009, at 01:15 UT. (Illustration © by the author)

- Using a black round-tipped brush tool, mark the positions of the brighter stars in and around the comet. Use a large pencil tip for the brightest, mentally using the clock system for position angle with respect to the comet at the center, along with an estimate in fractions of the radius to the field's edge. This cannot be done hastily. Take your time to place each individual bright star as accurately as possible, until you have enough field reference points to add in detail and the fainter stars. It will probably suit most observers not to go into too much detail in terms of the numbers and precise positioning of the fainter stars to save time and effort (Fig. 7.16).

Things to Note

Estimates of the brightness of a comet can be made by comparing it with defocused stars of known magnitude. While viewing a comet through the telescope eyepiece, be on the lookout for the shape and size of the false nucleus. Its degree of condensation

(DC) is measured on a scale that runs from 0 to 9; DC 0 equates to a totally diffuse, homogenous coma devoid of any condensation, whereas DC 9 is allocated to a comet with an obvious, distinctly star-like false nucleus. Particularly bright, active comets often show jets emanating from the false nucleus, along with subtle detail and structure nearby within the coma, such as shells. By using a polarizing filter it is possible to decrease the intensity of the coma so that only the brightest areas of the false nucleus and its surroundings are visible. Note the shape of the coma. It may be circular, symmetrical, fan-shaped, parabolic or irregular in outline; note also how diffuse it appears and the distribution of brightness.

Using steadily held binoculars, look out for waves, condensations, knots and disconnections in the comet's tail(s), and note their shape, orientation, length and any subtle coloration that may be visible.

In addition to the notes that ought to accompany every observational drawing — date, time (in UT), instrument, seeing, and so on, make a written note to clarify any interesting or unusual features. It is also important to note the orientation of the field of view depicted in your drawing, marking celestial north, south, east and west around the edge of the drawing. Special care must be taken to get this right if you're observing with a star diagonal. Another useful indication is to mark the drift direction of features across the field of view with an arrow at the edge of the circle.

Measuring Position Angle (PA)

Using an equatorial telescope, an illuminated astrometric reticle eyepiece and a homemade micrometer it's possible to derive fairly accurate position angles for any particular feature in a comet with respect to its nucleus — be it a tail, condensation or other measurable feature.

With the comet's nucleus at the center of the field of view, the position angle (PA) of the object being measured is stated in degrees, through east from north. So, the reticle must be as precisely aligned with the celestial cardinal points as possible, if accurate measurements of PA are to be taken. This is achieved by allowing a star near the celestial equator to drift across the field and rotating the reticle's linear diameter scale so that the star drifts parallel to it. Once aligned, a reticle attached to an astronomical telescope with a star diagonal (used by most SCT and MCT observers) will have north (0°) at the top of the field, east (90°) is at right, south (180°) at the bottom and west (270°) at the left of the field; PA is therefore measured clockwise from north. An astronomical telescope without a diagonal will have north (0°) at the bottom and east (90°) at the right, so PA must be measured in an anticlockwise direction from north.

A simple homemade micrometer can achieve results accurate enough to be of real scientific value. The author's homemade position angle micrometer is a CD stamped with a label printed with a clear graduated 360° scale around its circumference. Its center has been removed to accommodate the barrel of a 1.25-in. eyepiece, and it is firmly attached to a x2 Barlow lens. Used with the 12 mm reticle eyepiece, it delivers a magnification of x317. The Barlow is long enough to lift the micrometer clear of the ETX's 90° right-angle finder, and the CD's edge stops short of obscuring the finder's eyepiece, so the finder remains usable. The micrometer's pointer has been constructed from thin plastic sheet and attached to a parfocal ring fixed to the base of the reticle eyepiece. When the reticle eyepiece is loosened slightly from the Barlow, the pointer can be smoothly rotated around the 360° scale on the micrometer.

Given that the telescope's polar axis has been accurately aligned with the celestial pole, it's necessary to align the reticle eyepiece's linear diameter scale with east–west at the beginning of each observing session. Target a bright star near the celestial equator and allow it to drift across the field of view with the RA drive turned off. The aim is to rotate the reticle into a position where the target star drifts parallel with the reticle's linear diameter scale, with the scale precisely aligned east–west.

Using an astrometric eyepiece with a circumferential 360° scale, a rough determination of an object's PA can be gauged by positioning the comet's nucleus at the center of the reticle and estimating the angle made by the object being measured. Although this technique is not highly accurate, it provides a useful figure against which a more accurate micrometer measurement of PA can be compared. A semicircular PA scale in the reticle can also be employed.

To ensure an accurate measurement the micrometer's 360° scale must be aligned with the reticle. By moving the pointer on the micrometer to read 90° (east), the linear diameter scale ought to be precisely aligned with celestial east–west. Center the comet's nucleus and rotate the pointer clockwise so that the linear diameter scale in the reticle eyepiece is aligned with the object being measured. Read the position of the micrometer pointer on the 360° scale, which is equivalent to the PA of the object (Fig. 7.17).

Imaging

Bright comets of naked-eye magnitude or brighter can be reasonably well imaged using a regular digital compact or DSLR (digital single-lens reflex) camera, attached to a tripod and unguided, from a dark sky site. All other things being equal, the amount of detail you'll capture depends largely on the length of the exposure and the camera's ISO setting (the chip's light sensitivity). Some digital compact cameras have no means of making a time exposure, but those capable of taking exposures of a few seconds will be good enough to capture all of the naked-eye stars as well as the coma and perhaps hints of a tail, all without any appreciable star trailing. The large sensitive CCD chips within high-end digital compact cameras and DSLRs are quite capable of capturing the appearance of a bright comet and the subtleties of its tail in an exposure of a few tens of seconds. An undriven, wide-angle view of any part of the sky will begin to display noticeable star trailing after about 15 s, so it's best to set the ISO setting on the high side, bearing in mind that the higher the ISO, the 'grainier' the image.

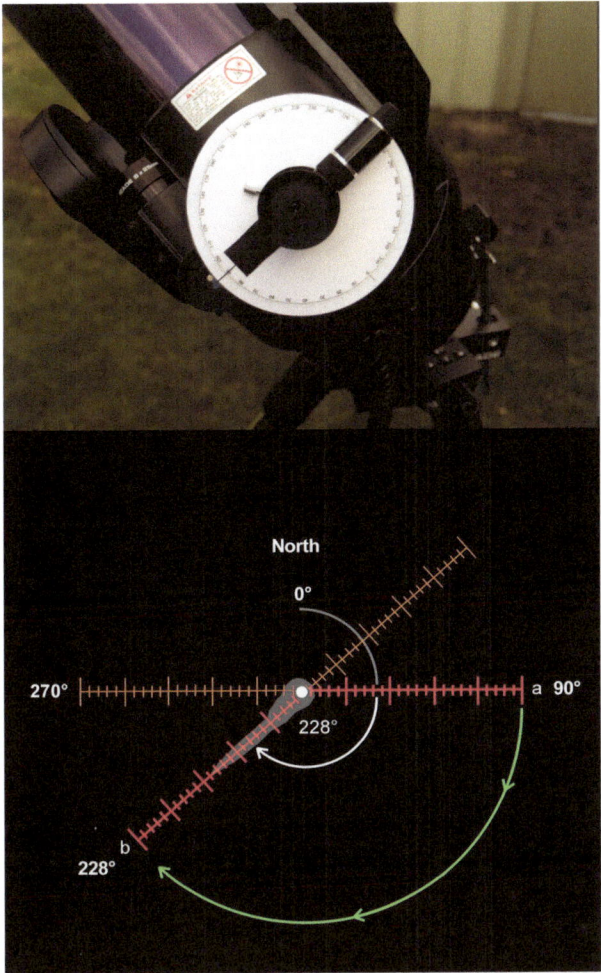

Fig. 7.17 *Top:* An illuminated reticle eyepiece, modified so that a surrounding 360° scale can be used to measure the position angle of a comet's tail. (Image by the author.) *Bottom:* With the comet's nucleus centered in the reticle, and the linear diameter scale aligned and pointing to 90° on the micrometer (**a**), the reticle is then rotated east so that it aligns with the comet's tail (or a component of the tails being measured). The micrometer is then read. This shows a measurement of a comet whose tail has a PA (position angle) of 228°. (Illustration © by the author)

A digital compact or DSLR camera piggybacked to a driven telescope, or mounted on its own equatorially driven platform, will enable much longer exposures to be made, allowing fainter background stars to be seen as well as revealing more color and detail within the coma and tail(s). If the platform is accurately

Fig. 7.18 A tripod-mounted digital compact camera, such as this Canon PowerShot SX130, can take undriven exposures of up to 30 s—ideal for capturing a bright comet in the sky above a suitable landscape. (Image © by the author)

aligned and driven, an SLR can be fitted with a telephoto lens for a closer view of the coma and detail within it (Fig. 7.18).

The best results are obtained by using an equatorially driven telescope as a large telephoto lens, the SLR being used at the telescope's prime focus (with the telescope minus its eyepiece and the camera minus its lens). Since a comet is in motion against the background stars, the best results during long exposures at relatively high magnifications will be obtained if the comet is kept at the center of the field of view. This can be achieved using a guide telescope or a flip mirror.

A regular modified webcam is perfectly capable of securing detailed images of the nuclear region of a bright comet, while an astronomical CCD camera capable of time exposures will enable advanced studies to be made. CCD imaging opens up a host of possibilities, including recording comets too faint to be viewed with the unaided eye, capturing detail within the coma of bright comets (including jets and shells), recording transient brightness fluctuations of the nucleus and making advanced astrometric measurements of a comet's position in the sky.

Chapter 8

Choosing the Right Equipment

Wide-angle images of comets are capable of showing the expanse, length of, and subtle detail within their tail(s). Even sky-spanning comets such as C/2006 P1 (McNaught), whose tail grew to 40° in length, was able to be captured in all its glory in a single image. Of course, no optical instrument has a wide enough field of view to provide a complete visual view of any comet whose tail stretches more than a hand's width across the sky (unless you look through your binoculars through the wrong end). Only the naked eye is capable of taking in the more extensive comets so that their scale can be appreciated (Fig. 8.1).

The Unaided Eye

As an astronomical tool, the naked eye tends to be less appreciated than it ought to be. For 100,000 years before the invention of the telescope, what the naked human eye saw was all that we thought existed in the heavens. As we've seen, a great deal was learned about comets, and even more was speculated upon, in the pre-telescopic era. Of course, far darker night skies prevailed in those dim and distant days: there were no streetlights, and there was no light pollution to speak of. People felt closer to nature, and the night skies were familiar places; the occasional unexpected eclipse, bright meteor or comet added mystery to the mix. These days, anyone lucky enough to experience a really dark sky site, far from the light pollution of the city, can experience what our ancestors took for granted (Fig. 8.2).

Regardless of the level of your regional light pollution—whether you live in a city, town or rural area—there are certain ways of using your eyes to achieve the maximum results. Far more complex and flexible than the most sophisticated digital

P. Grego, *Blazing a Ghostly Trail: ISON and Great Comets of the Past and Future*,
The Patrick Moore Practical Astronomy Series, DOI 10.1007/978-3-319-01775-4_8,
© Springer International Publishing Switzerland 2014

Fig. 8.1 C/2006 P1 (McNaught) imaged from Swifts Creek, Victoria, Australia, at around 10:10 p.m. on January 23, 2007. Taken at f/4, ISO 800, 20 s and ~24 mm with post-processing in Photoshop™ to bring out details. (Photo courtesy of Fir0002/Flagstaffotos.com.au)

Fig. 8.2 A comparison of the same area in Orion imaged with the same equipment and exposure time in dark skies and urban skies. (Image © by the author)

camera, the human eye consists of two compartments: the aqueous humor at the front of the eye and the larger vitreous humor, making up the bulk of the eyeball. Both cavities are filled with fluid and give the eyeball its strength and rigidity. Incoming light passes through the cornea, a transparent membrane covering the front of the eye, through the aqueous humor, then through a crystalline lens behind the pupil, across the vitreous humor, to finally impact onto the retina on the inside surface of the back of the eyeball.

Both cornea and lens focus the light, but the moment-by-moment fine-tuning is performed solely by the lens, whose degree of biconvexity (how 'fat' it is in the middle) is altered by muscles and fibers surrounding it. The amount of light passing through the lens is regulated by the iris, whose fine muscular fibers alter the diameter of the pupil, the aperture at its center. Light-sensitive cells in the retina convert the image—which, incidentally, has been turned upside-down by the lens—into electrical pulses that are sent to the brain and processed into an image we perceive as being the right way up.

Light from nearby objects consists of diverging rays that require a greater degree of refraction by the lens to focus properly. To do this, the ciliary muscles contract and the ciliary fibers slacken, allowing the lens to assume a more natural, fatter shape. As we age, our ability to focus objects decreases, as the lens becomes harder and less flexible. Light from very distant objects consists of near-parallel rays that don't require as much refraction by the lens to focus adequately. When viewing a distant object, the ciliary muscles surrounding the lens become relaxed, causing the ciliary fibers anchored around the edge of the lens to become taut. This pulls the lens into a thinner shape, allowing the image to be sharply focused onto the retina.

Our entire two-eyed field of view spans almost 180° horizontally and around half this vertically. It may surprise you to know that out of this wide panorama, the extent of distinct vision, where everything appears in sharp focus, takes up just a tiny portion of our entire field of view—around 2°—about the size of your thumbnail. When we look up at the night sky, we can only see a few stars at a time in sharp focus; beyond the very central part of our field of view, the rest of them are reduced to blurred smudges. But our perception of the night sky is far different.

Unable to rest upon a single static object and hold it at the center of our vision for any length of time at leisure, our eyes are continually scanning a scene to accumulate the maximum amount of information from it. Our eyes never can remain still for more than a few moments at most. Having evolved as wary little animals, fearful of predators or on the lookout for prey of our own, our eyes flit around involuntarily, assessing the scene continually, mainly for any signs of movement. This physiological phenomenon has direct consequences on attempting to observe detail in objects with a small apparent diameter.

Floaters are an annoyance experienced by many, and they tend to increase in number with age. They are the shadows cast onto the retina by the remnants of dead cells floating in the vitreous humor. Often annoying, they sometimes prevent viewing fine detail from time to time, but unless they are really severe they must be endured.

Our perception of objects and phenomena is also affected by the way that our brains are 'wired.' It is far more likely that a sudden, fast-moving object such as a meteor is noticed than a very slow-moving (or static) object of the same magnitude. A useful observing technique based on our ability to pick up movement can sometimes be used at the eyepiece, in order to coax a faint object into visibility. If a faint object, such as a dim comet, is known to lie somewhere within the field of view of an eyepiece and cannot be seen, it can sometimes pop into view when the image is made to wobble slightly—say, by rocking the telescope tube, or by some nifty fingerwork on the instrument's electronic controls. When searching for a faint fuzzy object, it might be seen immediately as it is made to move across the field of view, but once the telescope stops moving it can fade into invisibility once more.

Adapting to Darkness

Two types of light-sensitive cells are found in the retina. Cone cells at the center of the retina give us detailed color vision at the center of our field view, but they are only triggered by bright light. When observing a bright comet, you may see color in its tail(s).

Only the rod cells around the cones, some distance away from the center of the field of view, are triggered in dark conditions. The rods have less resolving power than the cones, so they deliver less detailed images. In addition, the rods cannot distinguish colors. A dim comet with a diffuse nucleus and coma may be difficult or impossible to see when it is looked at directly. But by using a technique called averted vision, observers shift their view a little way to the side of the actual position of the faint object, activating the light-sensitive rods, in order to see it better. The 'sweet spot'—where the faint object is best perceived—lies a little towards the observer's nose from where the direct vision is aimed. So, if you're viewing through the left eye, the faint object will be optimally perceived just to the right of the point you're actually looking at. It may all seem rather counterintuitive to anyone who has never tried averted vision before, but all seasoned visual observers use the technique. Incredibly, an observer with healthy eyesight may be able to see objects more than two magnitudes fainter using averted vision than with direct vision.

In very dark conditions, the eye's pupils dilate to their maximum size (around 8 mm across in a healthy young adult), allowing the maximum amount of light into the eye. In addition, a light-sensitive pigment called rhodopsin (sometimes called visual purple) is produced in the retina. Full dark adaptation never happens under urban conditions, since skyglow—caused by stray light from homes, businesses, industry and streetlights—illuminates dust and moisture in the air. Under dark country skies, stars of the 6th magnitude can be seen by anyone with average eyesight. This is enough for someone with good eyesight to pick out Uranus with the unaided eye. Sadly, the majority of the inhabitants of industrialized countries never get to appreciate a dark night sky. Skyglow drowns out faint stars and low-contrast objects such as the Milky Way. From a light-polluted city, often only the brightest objects can be seen. Observers near city centers are generally unable to see all but the brightest stars. Most comets, meteors, auroral glows, the zodiacal light and gegenschein are all astronomical phenomena denied to the urban astronomer's naked eye (Fig. 8.3).

A good degree of dark adaptation is required to telescopically observe most comets and to discern subtle detail within their comas and tails. Some observers don a pair of red goggles some time prior to their observing session. The eye is far less sensitive to red light, and it allows routine tasks to be performed while in the field while at the same time becoming dark-adapted. Maximum dark adaptation for an

Fig. 8.3 A comparison of the light-adapted and dark-adapted pupils. (Illustration © by the author)

urban observer is likely to be achieved after a quarter of an hour or so. To optimize dark adaptation, the observer should ensure that there are no local bright sources of light that may be looked at inadvertently. Car headlights or the sudden glare of an 'insecurity' light will instantly ruin any hard-fought-for dark adaptation. If illumination is required during the observing session—say, to consult a star chart or to sketch an object—put those red goggles back on or use a small red flashlight.

Optical Instruments

Although a great amount can be seen in the night skies with the naked eye alone, any form of optical aid will allow a closer, more detailed view of comets. Opera glasses, among the simplest optical devices, consist of a pair of basic, very short focus refractors, with single-element objectives and eye lenses. Since their lenses are considerably larger than the eye, they collect more light, reveal fainter objects and resolve more detail than the unaided eye; their magnification ranges between 2 and 5×, and as such their wide fields of view can accommodate all but the most sky-spanning comets. While opera glasses will show a comet in more detail than the unaided eye, they have one main optical shortcoming—that of chromatic aberration, where fringes of false color appear around the brightest objects such as the Moon and even the heads of very bright comets.

Binoculars

Binoculars have a more sophisticated optical configuration than opera glasses, and problems with chromatic aberration are minimized. These days, binoculars and telescopes of a tolerably high optical standard are widely available at prices to suit most people's budgets. Binoculars are a frequently underestimated tool in the amateur

Fig. 8.4 A pair of 10×25 roof prism binoculars. (Image © by the author)

astronomer's optical armory, for despite their often diminutive size, they pack a lot of punch. By using prisms to fold the light path, the length of binoculars is reduced to manageable dimensions. They usually have a right way up view so that they can be used for terrestrial viewing, and they are ideal for comet observation.

Porro prism binoculars are the most widely available type. Traditional Porro designs have the familiar W-shape, caused by the light path being folded inwards from widely spaced objective lenses towards the observer's eyes. Some small Porro prism binoculars have a U-shaped shell, and are more compact and easy to handle than traditional designs. They use inverted Porro prisms, so their objective lenses can be positioned closely adjacent to each other—in some cases closer together than the distance between the observer's eyes—but there is no real difference between the two types of Porro prism binoculars astronomically speaking.

Roof prism binoculars—so-called because of the particular shape of their prisms—are compact and lightweight. Most have a straight-through shape, appearing as though they are two small and ordinary refracting telescopes side by side. Inside the straight barrels of these binoculars, roof prisms fold the light path five times, producing a fairly long focal length instrument in a deceptively short body. Many monoculars are also of the roof prism design. Such instruments are very handy to carry around in a coat pocket or handbag and have ready at all times to snatch low power views of the skies (Fig. 8.4).

Fig. 8.5 A pair of 7×50 Porro prism binoculars. (Image © by the author)

Binoculars are by far the best starter instrument for a beginning amateur astronomer. Binoculars are labeled with two numbers, which indicate their magnification and aperture. For example, a pair of 7×50 binoculars (the best specification for general stargazing) delivers a magnification of 7 and has lenses of 50 mm diameter. Highly portable, such a small pair of binoculars can be used in corners of the backyard inaccessible to telescopes, and they can be carried to dark sky sites with ease. Binoculars are also capable of withstanding a few knocks due to their rugged construction (Fig. 8.5).

By holding binoculars up to the light and looking at the eye lenses from a distance, a small circle of light will be seen. The diameter of this circle called the exit pupil, and it can easily be calculated by dividing the size of the binoculars' objectives by their magnification. A pair of 12×50 binoculars have an exit pupil of 4 mm, while 7×50 s deliver an exit pupil of around 7 mm—about the size of a fully dark-adapted pupil. The larger the exit pupil, the better suited are binoculars to general astronomy since they deliver brighter views. However, binoculars with a larger exit pupil than your fully dilated pupils will not deliver all of their light to your retina, so they will underperform. The same principle applies to low-magnification telescope eyepieces. Light is wasted if their exit pupil is larger than your dilated pupil.

In binoculars, the shape and quality of the exit pupil is affected by the type of prisms used. BAK4 prisms used in good quality Porro prism binoculars produce bright circular exit pupils, while BK7 prisms used in many budget binoculars produce a distinctly square-shouldered exit pupil. Used in daylight, the two binoculars may perform equally well, since the eye's pupil is only taking in the central area of the exit pupil. Once light levels fall, and the pupil dilates, the shaded edge of the poor quality exit pupil affects image quality and brightness around the edge of the field of view. Thankfully, most binoculars are advertised stating the type of prisms they use, so an informed choice can be made.

With their wide fields of view, binoculars show comets to excellent effect. Indeed, they are the instrument of preference for many comet observers. Each year several comets are visible through 50-mm binoculars. Although they have the same light-gathering ability as a 50-mm refractor of equivalent focal length, the use of two eyes instead of one allows more detail to be discerned, and it makes for a more pleasant and relaxing experience, too.

For general stargazing, 7×50 or 10×50 binoculars are ideal. They deliver a wide field of view and a bright image, and have a magnification low enough to be able to smoothly and comfortably scan the skies for short periods. A field of view about 7° across is delivered by a pair of 7×50 s. That's an area of sky nearly 200 times larger than the full Moon. It's perfectly possible to enjoy general views of the night skies by simply holding a pair of lightweight, low power binoculars in the hands, without the need for any support, for several minutes at a time.

At the higher budgetary end but in the same size range there are image-stabilized (IS) binoculars—instruments whose prisms, linked to sophisticated internal electronics, actually correct for slight movements when hand-held, giving a shake-free view. The same sort of technology has been successfully adapted for most high-end digital cameras and vidcams. When used in hand-held mode, IS binoculars can deliver a one or two magnitude increase over simple hand-held binoculars; when both types are tripod-mounted, however, they perform just the same.

It is virtually impossible to enjoy a steady, hand-held binocular view using magnifications much greater than 15× (the upper limit of magnification for IS binoculars). Physical comfort while observing is essential, so some kind of support is required to keep binoculars steady and fixed on the object under scrutiny. When viewing objects at a high angle above the horizon through ordinary binoculars, the eyes must be aligned along the same optical plane as the binoculars. Using binoculars attached by a bracket to an ordinary photographic tripod, viewing any object higher than around 45° requires considerable craning of the neck—the higher the object, the more severe the craning and the more uncomfortable the observer's posture becomes. In addition, highly titled binoculars on a normal tripod may become somewhat out of balance, causing more problems. Some observers choose to lie in a garden lounger, bracing their elbows on the arms of the lounger (or using some form of independent mount) to achieve a comfortable, steady view of objects that are high in the sky.

Another solution is to use a specialized binocular parallelogram mount that allows the observer to stand at full height (or sit, or lie in one position) whatever the height of the object in the sky. Although they may look ungainly and impractical, parallelogram mounts are easy to use. The main arm of the mount has the binoculars at one end and a counterweight at the other; this is attached by pivot points to a parallel arm that keeps the binoculars aimed at the same spot of sky when the

Fig. 8.6 Tripod-mounted 15×70 Porro prism binoculars. (Image © by the author)

binoculars are raised or lowered—a handy facility to have if a number of people of different heights wish to observe in comfort. Using a parallelogram mount it is possible to access a large area of the sky without greatly altering position.

Giant binoculars—those with objectives larger than 70 mm—certainly require a firm, steady mount to observe with. Most of these instruments are manufactured with a central bar, connected to both tubes, on which is attached an adjustable mounting bracket. The best form of mount for these kinds of binoculars is a heavy-duty parallelogram mount, or at the very least a heavy-duty fully adjustable tripod. Some high-end big binoculars manufactured specifically for astronomy and comet-hunting are tripod-mounted and have eyepieces angled by 45° for viewing comfort (Figs. 8.6 and 8.7).

It may be tempting to purchase a pair of giant zoom binoculars, say, a pair of 20 to 100×70 binoculars, but unless the binoculars are a premium brand, it's probably best to avoid these, since any slight defects in build quality will become exaggerated when higher magnifications are used to the point where the brain is incapable of assembling a single image from the two overlapping fields of view (Table 8.1).

Fig. 8.7 Tripod-mounted 25 × 100 Porro prism binoculars. (Image © by the author)

Table 8.1 Typical binocular fields of view

Instrument	Field of view	Exit pupil (mm)
2 × 40 (opera glasses)	30°	20
8 × 21	6.5°	2.6
10 × 25	6.5°	2.5
8 × 42	6.3°	5.3
7 × 50	6.3°	7.1
10 × 50	5.8°	5
12 × 50	5.4°	4.2
15 × 70	4.1°	4.6
20 × 80	3.1°	4
25 × 100	3.1°	4

Many amateur astronomers begin with a small, unsophisticated telescope and upgrade to larger instruments as their budget allows. Small telescopes have a number of advantages. They are lightweight, portable, and to some extent expendable. Small telescopes are often sold with a selection of two or three eyepieces. The maximum usable magnification for any telescope is around twice the objective's size in millimeters, e. g., 100× for a 50-mm telescope. Higher magnifications than this will not produce a better, clearer image.

Although the optical quality of a small budget telescope's objective lens or mirror may be good, the eyepieces provided may not be good enough to get the best performance from the instrument. Star diagonals may have such poor quality mirrors that they are unusable, and the Barlow lens—an accessory that doubles or triples the magnification delivered by any eyepiece—is not likely to improve the view. By using good-quality eyepieces that deliver low to moderate magnifications, the instrument will earn its keep, and the observer will enjoy some lovely views. It is worth obtaining one or two good quality 1.25-in. Plössl eyepieces, along with a 0.96–1.25 in. eyepiece adapter if required.

Refractors

Refracting telescopes consist of a tube with an objective lens at one end, which collects and focuses light, and an eyepiece at the other end, which magnifies the focused image. The focal length of a refractor is expressed as a multiple of the lens diameter. A 100 mm diameter objective of f/10 will have a focal length of 1,000 mm. The most basic type of refractor has a single objective lens and a single eyepiece lens. Reflectors require very long focal lengths to combat the effects of false color, known as chromatic aberration, caused by the different degrees to which the constituent colors of light are refracted by the objective lens. Through such a basic refractor, a bright cometary head may appear to be surrounded by fringes of colored light (Figs. 8.8 and 8.9).

Most of the effects of chromatic aberration are overcome in refractors that use an objective lens consisting of two specially shaped lenses sandwiched together. Achromatic objectives focus the different wavelengths of visible light near to a single point. Achromatic refractors don't entirely eliminate false color. Residual chromatic aberration is particularly noticeable in budget value short-focus refractors of f/8 and below. Although the edges of bright objects display a tinge of violet, this is aesthetically unobtrusive to most observers, and the unwanted hue can be eliminated with a minus-violet filter, which fits into the eyepiece. It is also possible to correct color aberrations in refractors using a special correcting lens that fits into the focuser, and although this option can be expensive, it is less expensive than the cost of buying a fully fledged apochromatic instrument.

Fig. 8.8 An 80-mm rich-field achromat. (Image © by the author)

Apochromats deliver images free from chromatic aberration by using special glass in their two- or three-element objective lenses, serving to focus all wavelengths of visible light into as near a single point as possible. Capable of delivering near-perfect, high-contrast images, apochromats are among the most desirable telescopes for astronomical observation. On average, though, they are ten times costlier than achromats of a similar-size.

Reflectors

By using mirrors to collect and focus light, reflecting telescopes produce images free from the effects of chromatic aberration. Newtonian reflectors are the most commonly used type of reflector. They use a concave parabolic primary mirror at the base of the telescope tube, the converging light from which is reflected out of the side of the tube and into the eyepiece by a small flat mirror near the top of the tube. This smaller secondary mirror gets in the way of the light path entering the telescope, but it only prevents light from reaching a small portion of the central area

Fig. 8.9 A go-to 4-in. (102-mm) f/6.6 rich-field achromat. (Celestron NexStar 102 SLT). (Image © by the author)

of the primary mirror. Therefore, a Newtonian of a certain aperture has less of a light-gathering surface than a refractor with an objective lens of equivalent diameter.

Because secondary mirrors and their supporting structures intrude into the light path, they cause some distortion of the image. Spikes can be seen around stars, caused by the scattering of light. This phenomenon, known as diffraction, causes a certain amount of contrast in an image to be lost—a property most valuable in being able to discern subtle lunar and planetary detail.

The effects of diffraction in Newtonian reflectors become less severe with longer focal lengths, since they have smaller secondary mirrors because the diameter of the converging cone of light intercepted from the primary mirror will be smaller. A 150-mm f/8 Newtonian reflector—one of the most commonly found Newtonian specifications—is capable of giving pleasing all around celestial views, and can be used at high magnifications for studies of structure visible in the coma and near the nucleus. For detailed high magnification observation, longer focal lengths will perform better. A well-collimated long focal length Newtonian reflector, say one of f/10 or greater with high grade optics, is capable of delivering views that are of apochromatic quality. The chief disadvantage of long focal length Newtonians is the length of their tube—a 250 mm f/10 will have a focal length of 2.5 m. Such an

Fig. 8.10 The author uses his 17.5-in. (445-mm) f/4.5 Newtonian (Dobson-mounted). (Image © by the author)

instrument will require a substantial, sturdy tube and mount, while a 250 mm f/5 Newtonian is quite capable of being used as a portable instrument on a Dobsonian (a basic up-down, side-to-side, altazimuth) mount (Figs. 8.10 and 8.11).

Eminently suited to high-powered observation, Cassegrain reflectors have a reputation for being difficult instruments to make, maintain, adjust and collimate, and their use has never been very widespread among amateur astronomers. Cassegrains have a concave secondary mirror that reflects light onto a convex central secondary mirror, which in turn reflects the light back down the tube, through a hole in the primary and into the eyepiece. With their twice-folded light paths, Cassegrains pack a lot of focal length into their tube (usually a skeleton tube, to cut the weight of the instrument and to minimize the effects of tube currents). Cassegrains are prone to the twin optical aberrations of astigmatism and field curvature, and for these reasons they are usually found in very high focal lengths, from f/15 to f/25.

Catadioptrics

Catadioptric telescopes use both mirrors and lenses to collect and focus light. Schmidt-Cassegrain telescopes (SCTs) are the most popular catadioptric used by modern amateur astronomers and Solar System observers. A hybrid of the Schmidt camera and Cassegrain telescope design, SCTs are closed systems consisting of a

Fig. 8.11 The author's 12-in. (300-mm) f/4.5 Newtonian, mounted as a Dobsonian. The mirror and Dobsonian were constructed by the author. (Image © by the author)

perforated primary mirror and a secondary mirror fixed to the inside of a whole aperture lens. Known as a correcting plate, the lens appears thin and flat, but it is very minutely figured so that the light passing through it is slightly bent so that the rays are reflected to a focus by the concave spherical primary mirror. Light is reflected from the convex secondary through the central perforation in the primary mirror, into the eyepiece. In common with SCTs, Schmidt-Newtonian telescopes (SNTs) have a full aperture frontal corrector plate, and their secondary mirror reflects light out of the side of the tube into the eyepiece (Fig. 8.12).

With such a relatively large central obstruction, diffraction within SCT and SNT systems can very slightly reduce the quality of the image and lessen contrast when viewing detail on the Moon and planets, a point often overemphasized by zealous telescope purists. In truth, a well-collimated SCT or SNT with good optics will produce really excellent views of Solar System objects, including comets. Indeed, many of the world's top observers and astronomical imagers use SCTs.

SCTs can easily accommodate a wide range of useful accessories, such as filter wheels, cameras, webcams, digicams, vidcams and astronomical CCD cameras. When viewing objects directly through the eyepiece, an SCT produces a 'traditional' inverted astronomical view. For ease of use, SCT users often use a mirror star

Fig. 8.12 The author's 8-in. (200-mm) f/10 SCT, (Meade LX90), a go-to telescope. (Image © by the author)

diagonal, which gives a reverse image of the already inverted image. If a 'right way up' image is required (one that is orientated the same way as the naked-eye view), an erecting prism, 45° prism or regular prism diagonal will do the job (Figs. 8.13 and 8.14).

Maksutov-Cassegrain telescopes (MCTs) use a spherical primary mirror and a deeply curved spherical meniscus lens at the front of the tube to correct for spherical aberration. The secondary mirror itself is a small spot that has been aluminized directly onto the interior of the meniscus. In their general appearance, MCTs resemble SCTs in many ways, but with their longer focal lengths, their performance on comets is generally better. They offer near-apochromatic quality, high resolution, high contrast views without any appreciable chromatic aberration.

Catadioptrics are sealed systems, and the only optical surface directly exposed to the outside world is the exterior of the corrector plate or meniscus. Great care must be taken when getting rid of any debris that accumulates on this component, as it is often coated with contrast-enhancing material. SCTs can come out of collimation over time, and this can usually be made right again by means of three small collimation screws on the exterior housing of the secondary holder. Being closed systems, catadioptrics require a certain cool down time to equalize

Fig. 8.13 Meade f/6.3 focal reducer for the 8-in. (200-mm) SCT (Meade LX90), enabling wider fields to be viewed while improving edge-of-field correction and reducing astro-imaging exposure times. (Image © by the author)

Fig. 8.14 An 8-in. (200-mm) SCT (Meade LX90) with 2-in. diagonal. (Image © by the author)

their internal temperature and that of their optics with the outside air. This can be sped up by means of an internal fan in some larger catadioptrics.

Treated with care, catadioptrics require much less maintenance than reflectors, whose mirrors are open to the elements. Their mirrors can last for decades without requiring re-aluminizing. Perhaps the biggest downside of these versatile instruments is that their corrector plates or menisci require some kind of protection from dew, as it is liable to accumulate very fast given certain conditions, putting an end to any observing session for a while. A dew cap or electrical dew zapper (a very mild heating element that wraps around the edge of the corrector plate housing) is an essential requirement if you want to follow comets at any time of the year.

Eyepieces and Magnification

New observers often overlook the fact that the eyepiece is as important to the performance of a telescope as its objective lens or primary mirror. A poor-quality eyepiece will deliver a substandard image. At best this will frustrate the observer and may act as a deterrent to further observational astronomy. Many budget instruments are provided with inadequate eyepieces—a great pity, since the telescope optics themselves may be of excellent quality.

Eyepieces serve to magnify the image; to calculate the magnification provided by a given eyepiece, simply divide the focal length of the telescope by the focal length of the eyepiece. For example, a 9 mm eyepiece will give a magnification of 133× in a telescope with a focal length of 1,200 mm (1,200/9 = 133).

Magnification is important, but it shouldn't be the sole factor in choosing an eyepiece. Apparent field of view is the width of the image visible through an eyepiece, expressed in degrees, when the eye is at the optimum distance from it. A larger apparent field of view tends to enhance the observing experience. Apparent fields of 40° or less give the impression of peering up the shaft of a deep well at the night sky above. Apparent fields of view over 70° are often described as a 'spacewalk' experience—so wide that one's eye can't take in the whole field at once—but such an experience doesn't come cheap.

Three good-quality eyepieces giving different magnifications should be enough to satisfy the basic needs of the observer:

1. A low power eyepiece with a fairly wide actual field of view, enough to take in the comas of most comets.
2. An eyepiece of medium power (up to 100×) for more detailed observation of the coma and nucleus.
3. A high power eyepiece. The highest reasonable magnification to use on nights of average seeing can be found by doubling a telescope's aperture in millimeters, e. g., 300× for a 150-mm reflector. This eyepiece can be used for observing fine detail, if present (Fig. 8.15).

Fig. 8.15 A selection of wide-field 2-in. eyepieces ideal for use on a 12-in. (300-mm) f/4.5 Newtonian: 30 mm ultrawide (80° apparent field), magnification 45×; 24-mm Hyperion super-wide (68° apparent field), magnification 56×; 15 mm ulta-wide (80° apparent field), magnification 90×. (Image © by the author)

Of the three eyepiece barrel diameters generally available—0.965, 1.25, and 2 in.—1.25-in. ones are the most commonly used. An increasing number of focusers are fashioned with an adaptor to accept both 1.25 and 2 in. diameter eyepieces. The 2-in. eyepieces are usually of the low magnification, wide angle variety (Fig. 8.16).

Light-Gathering Ability and Resolving Power

The larger the telescope lens, the greater its light-gathering ability. Light from a point source, such as an asteroid, is concentrated into a focused point, so it appears brighter and easier to see through a large telescope than it does through a smaller telescope. Viewed through the eyepiece, the brightness of a point source is proportional to the area of the telescope's objective lens (or mirror). A telescope's light-gathering ability is often expressed in terms of the magnitude of the faintest stars

Fig. 8.16 A comparison of 2, 1.25, and 0.965 in. eyepieces. (Image © by the author)

that it will reveal under ideal conditions. The limiting magnitude—the magnitude of the faintest star that can be seen with a given instrument—is the most convenient and meaningful measure of its light-gathering power.

Aperture (mm)	Limiting magnitude
Unaided eye	6.5
25	9.1
60	11.0
100	12.1
150	13.0
200	13.6
250	14.1
300	14.5

To find the limiting magnitude M visible through a telescope:

$$M = 6.5 - 5 \log d + 5 \log D$$

where d is the dark-adapted pupil's diameter (in millimeters), D is the instrument's aperture (in millimeters) and 6.5 is limiting magnitude for stars visible with the unaided eye. d amounts to around 7.5 mm in a healthy young adult in really dark conditions, but will be considerably smaller in light-polluted urban conditions (and so, too, will the limiting magnitude of the unaided eye). The table above is based on an ideal combination of seeing, observer and instrument.

Resolving power, the ability to discern fine detail, increases with telescope aperture. There are a number of factors that determine how high a magnification it is reasonable to use. Chief among these is the steadiness of the column of Earth's atmosphere through which the light from a celestial object passes before reaching the telescope, known to astronomers as 'seeing.' On nights of good seeing, the maximum usable magnification is around twice the diameter of the aperture (measured in millimeters).

Aperture (mm)	Max magnification	Resolution (arcs)
Unaided eye	1	20"
25	50	4.6"
60	120	2.0"
100	200	1.2"
150	300	0.8"
200	400	0.6"
250	500	0.5"
300	600	0.4"

The resolving power for various instruments given in the table above represents a practical guide, rather than the theoretical predicted by the Dawes' Limit equation. To find the Dawes resolving limit R (in arcseconds) for a telescope of aperture D (in millimeters):

$$R = 116 / D$$

Seeing

After passing smoothly and uneventfully through the vacuum of space, light from celestial objects is subjected to a buffeting by Earth's atmosphere before finally encountering the astronomer's telescope. Turbulence in the atmosphere affects the quality of the image of any astronomical object, the quality astronomers call seeing. Warm air, released from a landscape cooling down after a warm day, or from houses and industry, interacts with the colder night air, producing a turbulent mix of different densities and refractive properties.

The flow of the air over the landscape can be disrupted by hills and buildings to form eddies that can further mix and degrade the seeing. Stars nearer the horizon, whose light passes through a greater amount of atmosphere, are most prone to scintillation. Being extended objects rather than point sources, the Moon and planets don't usually appear to twinkle, even when they're low down. But that's not to say that they are immune from the effects of turbulence. Through the eyepiece, atmospheric turbulence can play havoc with lunar and planetary observation, sometimes being so bad as to render observations futile.

A simple scale of seeing was devised by the great planetary observer Eugene Antoniadi (1870–1944), which could be used by observers to describe the quality of the atmosphere at the time of each observation:

AI   Perfect, image exceptionally steady
AII   Good, generally steady image with occasional undulations
AIII   Moderate, frequent undulations with some moments of calm
AIV   Poor, undulations constant
AV   Appalling, hardly worth observing

On the basis of observations made through a 12 mm refractor, William H. Pickering (1858–1938) of Harvard Observatory devised a more scientifically based scale of seeing. The steadiness of the atmosphere was gauged by the visible detail and extent of the Airy disk (the small circle of light into which the light from a point source, such as a star, is concentrated) and diffraction patterns (artifacts surrounding a star's image caused by the instrument's optics). Users of the scale rarely go to the lengths of actually assessing the diffraction patterns visible, since regular observers are quite capable of accurately judging the quality of an image and assigning its position on a scale of 1–10. In the scale, 1–2 is very poor, 3–4 is poor, 5 is moderate, 6–7 is good, 7–8 very good, and 8–10 is excellent.

P1   Star image is usually about twice the diameter of the third diffraction ring if the ring could be seen; star image 13″ in diameter.
P2   Image occasionally twice the diameter of the third ring (13″).
P3   Image about the same diameter as the third ring (6.7″) and brighter at the center.
P4   The central Airy diffraction disk often visible; arcs of diffraction rings sometimes seen on brighter stars.
P5   Airy disk always visible; arcs frequently seen on brighter stars.
P6   Airy disk always visible; short arcs constantly seen.
P7   Disk sometimes sharply defined; diffraction rings seen as long arcs or complete circles.
P8   Disk always sharply defined rings seen as long arcs or complete circles, but always in motion.
P9   The inner diffraction ring is stationary. Outer rings momentarily stationary.
P10   The complete diffraction pattern is stationary (Fig. 8.17).

Choosing Eyepieces

Basic Eyepieces

Huygenian-, Ramsden-, and Kellner-type eyepieces are all very old designs that deliver small apparent fields of view. They are usually provided with small budget instruments, and their performance may disappoint. Huygenians are suitable for use

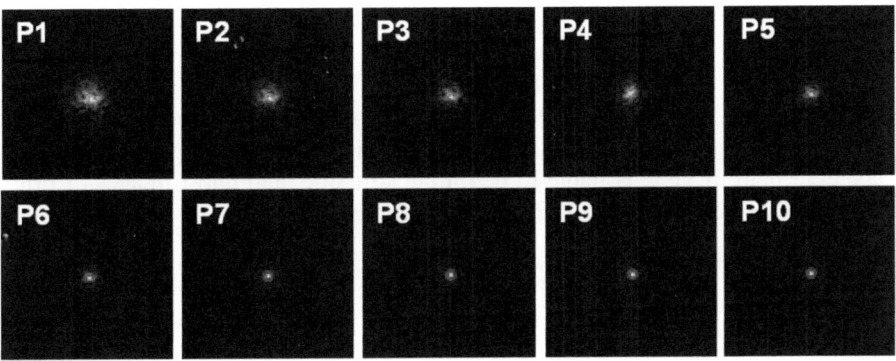

Fig. 8.17 The Pickering scale of seeing. (Illustration © by the author)

only with long focal length instruments of f/15 or more, and give apparent fields of just 30° or less. Ramsden eyepieces deliver flatter fields than Huygenians, but they produce more chromatic aberration. Kellner eyepieces give better contrast but tend to produce annoying internal ghosting when viewing bright objects such as the Moon and planets. Kellners perform adequately at focal lengths longer than 15 mm and have good eye relief, but shorter focal length versions are blurred at the edge of the field of view and introduce unsightly chromatic effects. They are sometimes referred to as MA (modified achromatic) eyepieces. Both the Ramsden and Kellner designs have focal planes in front of the field lens, meaning that any tiny specks of dust that happen to land on the field lens are brought into view. Neither Huygenian, Ramsden nor Kellner eyepieces meet the demands of the modern Solar System observer.

Quality Eyepieces

Orthoscopic eyepieces produce a flat, aberration-free field, and they deliver very good high contrast views of the Moon and planets. Their apparent field of view is around 50°, and they have a reasonable degree of eye relief.

Probably the most popular eyepiece among observers, Plössls have minimal internal reflections, good color correction and an apparent field of view around 50° that is flat and sharp even up to the edges. Their main disadvantage is that of poor eye relief with lower focal length eyepieces, requiring the eye to be very close to the lens to take in the whole field of view of a high magnification scene. Longer focal length Plössls give acceptable eye relief, and there are special long eye relief versions with large eye lenses that allow wearers of eyeglasses to observe in comfort.

Developed in the nineteenth century by Carl Zeiss, Monocentric eyepieces are an old design, but they probably offer the best performance for high magnification observing. Although they only deliver a tunnel-like 30° field of view, Monocentric optics are virtually scatter-free, producing high contrast images whose quality cannot be excelled by any other eyepiece.

An eyepiece with a somewhat wider apparent field of view, the Erfle, gives a low magnification field with good color correction. Erfles perform at their best when used with long focal length telescopes. Definition at the edge of their 70° wide apparent field of view is liable to be on the soft side, and ghost images of bright subjects may be visible due to internal reflections.

Wide Angle Eyepieces

Introduced in recent decades, modern wide field and ultra wide field eyepieces such as the Meade Ultrawides, Celestron Axioms, Vixen Lanthanum Superwides and TeleVue Radians, Panoptics and Naglers deliver well corrected images with very large apparent fields of view. Larger fields of view are easier to use on an undriven telescope, because the subject stays in the field for longer, and telescope's aim doesn't require adjusting as often. Naglers, with their 80° plus apparent fields, are splendid eyepieces that deliver breathtaking wide angle views. The longer focal length Naglers are big and very heavy. The 13-mm Nagler is 135-mm tall and weighs a hefty 0.7 kg. Switching between eyepieces of this sort and regular eyepieces almost certainly requires the telescope to be rebalanced each time. The recent introduction of mega-wide eyepieces with 100° (plus) apparent fields of view, such as the TeleVue Ethos, have made the 'spacewalk' experience seem even more real, but they are by no means inexpensive eyepieces.

Zoom Eyepieces

Zoom eyepieces are no longer just a novelty item. A number of quality zooms are on the market, allowing the observer to change between focal lengths of, say, 8 to 24 mm, at the twist of the eyepiece barrel. Their apparent field of view narrows from a generous 60° at 8 mm to a somewhat miserly 40° at 24 mm, but if you can accept this restriction, a single zoom eyepiece can replace a drawer full of regular Plössl eyepieces. Although most zoom eyepieces require a little refocusing after changing their focal length, the Nagler zoom remains in focus throughout its 3–6 mm range.

Zoom eyepieces go well with the binocular viewer—a device that splits the single beam of light from the objective and diverts it into two identical eyepieces. Because of their design, most binocular viewers will work only on a refractor or an SCT. It may not be possible to focus one using a Newtonian, since it requires a long light path. Binocular viewers can only be used with two identical eyepieces, and these ought to be of a focal length shorter than about 25 mm in order to prevent image degradation around the edge of the field of view. A binocular viewer adds a

Table 8.2 Actual and apparent fields for different eyepieces

Magnification	Apparent field	Actual field
25	50°/80°	2°/3.2°
50	50°/80°	1°/1.6°
75	50°/80°	41'/1.1°
100	50°/80°	30'/48'
150	50°/80°	19'/32'
200	50°/80°	15'/24'
300	50°/80°	10'/16'

Table 8.3 Actual examples based on a 200 mm f/10 SCT

Eyepiece	Magnification	Apparent field	Actual field
Meade Series 4000 (56 mm)	36	52°	1.4°
Celestron Ultima (42 mm)	48	36°	45'
Meade Series 4000 SWA (32 mm)	63	67°	1.1°
TeleVue Nagler (31 mm)	64	82°	1.3°
Meade Series 3000 (25 mm)	80	50°	38'
Celestron Axiom (19 mm)	105	70°	40'
TeleVue Nagler (17 mm)	118	82°	42'
TMB Monocentric (14 mm)	143	30°	13'
TeleVue Ethos (13 mm)	154	100°	1.1°
Baader Hyperion (10 mm)	200	68°	34'
TeleVue Nagler (7 mm)	286	82°	17'
Celestron Ultima (5 mm)	400	50°	5'
TMB Monocentric (4 mm)	500	30°	4'

sense of perspective to observations, producing an almost 3-D quality to the scene; certainly, there's something very special about allowing both sides of your brain to simultaneously appreciate a celestial vista. It has also been demonstrated that viewing with both eyes enables much finer detail to be discerned (Tables 8.2 and 8.3).

Barlow Lenses and Focal Reducers

An extra lens can be placed in the light path before it enters the eyepiece in order to extend or shorten the effective focal length of the telescope, altering the magnification delivered by any given eyepiece.

Fig. 8.18 A 1.25 and 2-in. Barlow lens. (Image © by the author)

A negative lens called a Barlow lens effectively increases a telescope's focal length—usually doubling or tripling it, hence doubling or tripling the magnification delivered by an eyepiece. Barlow lenses are commonly used, and it is sometimes better to use one with any particular eyepiece if higher magnification is required, rather than switching to a shorter focal length eyepiece with poorer eye relief (Fig. 8.18).

A focal reducer/corrector is a positive lens frequently employed on SCTs. The most common type is one that reduces the focal length of a typical f/10 SCT to f/6.3. This reduces the magnification of any given eyepiece by 37 %, producing an actual field of view some 160 % wider and 2.5 times greater in area. A reducer/corrector is ideal for taking in the wide fields of view required for comet hunting and observing. Reducers/correctors also flatten the field of view to the edge, making them a better option for use with a given eyepiece than a lower magnification eyepiece, which may show a degree of edge distortion when used without the reducer/corrector. Shorter astrophotography exposure times are possible using reducer/correctors, and they deliver a larger field on the CCD chip.

Telescope Mounts

Telescope mounts require a high degree of rigidity and stability. Altazimuth mounts allow the telescope to be moved up and down and swung from side to side, and the most popular type is the Dobsonian. Simple and easy to use, Dobsonians consist of a box with an azimuth bearing that rests on the ground; this carries another box (and the telescope tube) with an altitude bearing is at its center of balance. Low friction materials such as polythene, Teflon and Formica are used for the load-bearing surfaces, which allows easy fingertip control. Lightweight structural materials such as plywood make Dobsonian-mounted instruments very portable.

Earth's rotation causes every celestial object to make a complete circuit around the celestial pole every 24 h. An object located near the celestial equator will appear to move westwards at the rate of one quarter of a degree per minute. This may not seem like much—indeed, the motion can't be detected with the unaided eye—but once the sky is magnified through the telescope eyepiece, it becomes apparent. A small telescope on an undriven altazimuth mount can only be used to observe comfortably using lower powers, up to a maximum of 100×. Much higher magnifications require adjusting too frequently to keep an object in the field of view long enough for it to be properly observed.

An eyepiece with a 50° apparent field giving a magnification of 100× will take in an actual field of view about half a degree wide—about enough to fit in the whole Moon. Without a drive, an object will appear to drift from the center of the field to the edge in less than a minute. Making the light adjustments needed to keep an object centered with the field of view can be a difficult task using telescopes mounted on photographic tripods or simple altazimuth mounts with clamp-friction bearings.

There are two solutions available to make a Dobsonian-mounted telescope follow the drift of the stars. The Poncet mount can be retrofitted onto virtually any Dobsonian. This serves as a part-revolving base upon which the entire Dobsonian sits. Its drive system, part of a conic section aimed at the pole, is capable of maintaining the instrument in equatorial mode for extended periods of time, enabling comfortable viewing and astro-imaging.

It is possible to self-manufacture such a mount (plans are available online), but there are several manufacturers producing top quality gear. Alternatively, some commercially available Dobsonians are provided with electronic drives attached to both azimuth and altitude axes; controlled using a computerized handset, this high-tech Dobsonian becomes an equatorial with a go-to facility. Astroimaging is possible with such an instrument, but because of the altazimuth nature of the drive there is noticeable field rotation on long exposures (this is not the case with the Poncet mount). It is possible to retrofit a basic Dobsonian with an altazimuth drive. A commercial kit to do this is available from a number of sources.

Equatorial mounts allow the telescope to follow celestial objects as Earth rotates. A properly set up equatorial mount, with its polar axis aligned with the celestial pole, allows the observer to keep a celestial object centered in the eyepiece with ease.

Even if the mount has no motorized drive, a light push on the tube or a slight twist of the slow motion knob on the RA axis from time to time will keep the object in view far easier than the pushing and pulling required of an altazimuth mount. A motor drive makes for the most hassle-free viewing experience.

Two types of equatorial mount are in widespread use. German equatorial mounts are sturdy and flexible, often used to mount refractors and reflectors. The entire sky above the horizon, including the celestial pole, is accessible with a German equatorial mounted telescope. SCTs are usually mounted between the arms of heavy-duty fork mounts. Many SCTs are capable of accurate computer-controlled equatorial tracking in altazimuth mode, both axes being adjusted continually. For greater tracking accuracy the whole fork mount can be tilted to point to the pole, requiring only one axis to be driven to keep an object centered in the field. Fork-mounted instruments are often denied a view of a small region near the celestial pole, since any large eyepieces or accessories such as binocular viewers or CCD cameras will not permit the telescope to move between the fork and the very base of the mount.

We've presented a general overview of the sort of gear available to the amateur astronomer, with emphasis on the sort of equipment that is likely to be of use to the comet observer. Importantly, your enjoyment of comets oughtn't to be dependent on how much equipment you have or how expensive it might be. Many memorable cometary views have been made with the unaided eye and through binoculars. Sure, large telescopes are capable of showing great detail (if present) around a comet's nucleus, such as shells or jets of material, but that's a little like admiring just a small section of an Impressionist painting. In the next section we'll find out how to use your equipment to locate (or even discover) comets and how best to record them.

Chapter 9

ISON and Future Comets

New comets are discovered pretty frequently—some by fortunate but vigilant amateurs, others by professional astronomers using automated search programs. Often these discoveries brighten and become respectable binocular or even naked-eye objects at their peak. Then there are the periodic comets, 290 of which are known; their orbits are within the Solar System and their behavior can be predicted with a good degree of accuracy for many perihelia to come.

First, a couple of the new discoveries—non-periodic comets—that should become bright enough to be enjoyed in the coming years. The list is too big to include everything here, but there are many resources that can be consulted, both online and in print, in order to find out precisely what is observable, when and where.

C/2012S1 (ISON), 2013–2014

Discovered on September 21, 2012, C/2012S1 (ISON) was soon found to be new comet, arriving fresh from the Oort Cloud. Shortly after its discovery it was realized that the comet would come incredibly close to the Sun in November 2013, promising to blaze brightly during late 2013 and early 2014. Vitali Nevski, co-discoverer of Comet ISON (with Aryton Novichonok) of the International Scientific Optical Network (ISON) Survey from near Kislovodsk, Russia, described the excitement of discovery:

> It was first unstable night after ten ones of hard work. The endless clouds disturbed the sky. That night Artyom and I decided to look through early discovered asteroids with the most interesting orbits. The sky became clean literally for half an hour before dawn and it was decided to launch continuous random search platform on the border of Twins and Cancer.

P. Grego, *Blazing a Ghostly Trail: ISON and Great Comets of the Past and Future*,
The Patrick Moore Practical Astronomy Series, DOI 10.1007/978-3-319-01775-4_9,
© Springer International Publishing Switzerland 2014

After the series ended Artyom went to have a rest, and I loaded platform in CoLiTec [automatic detection software for asteroids and comets] and remained to wait the processing completion learning earlier discovered astrometry of asteroids.

When I started the programme of completing results the first thing I noticed among the third position objects was the bright one with unusual movement. The object was not identified with the MPC base. The unusual thing was rather slow movement relative to the asteroids of the main zone in this region. At that moment I started to realize that so slow object did not belong to the asteroid zone and could be situated only far away behind Jupiter's orbit. My heart missed a beat. Is it really a comet? There was nothing to do except to deposit the astrometric data on the site NEOCP for acknowledgement.

In the evening I asked Artyom to book the time for observing of the object with the 1.5-m telescope of Maidanak Observatory to be sure of the cometary nature of the object. Artyom and I looked through old materials, analyzed possible orbits of a new object, and by then there appeared some new astrometry points. About 4 O'clock in the morning we saw the pictures—there was no doubt that it was a comet, a classical compact one measuring 9×11 arcs, similar to long-period comets. We immediately sent a message to IAU and awaited results. We were afraid that the object had already been discovered. But the most interesting thing was that the orbit of the comet started to appear with the introduction of new points of astrometry and the analysis showed perihelion near the Sun, so the brightness of the comet in the maximum could reach the full Moon.

Finally on October 24 there was a circular and it was hard to believe that perihelion of the comet was 0.01 au and at the maximum brightness could reach such negative magnitudes that the comet could become the comet of the century! (Fig. 9.1).

At the time of its discovery Comet ISON was located in northeastern Cancer and a dim magnitude 18.8, some 6.29 au (941 million kilometers) away from the Sun, just outside the orbit of Jupiter. Initially visible in the morning skies but only accessible to those with very large telescopes and advanced imaging equipment, it described a loop, heading east into neighboring Gemini and passing near the star Castor in mid-January 2013. On January 30 NASA's Swift spacecraft observed the comet in ultraviolet light, discovering that its nucleus was releasing 850 tons of dust per second—and the comet was then 2.7 au from the Sun. It was estimated that the nucleus of Comet ISON was around 5 km across. Its production of water vapor was on the low side, and the driving force behind the jetting of material seemed to be the sublimation of carbon dioxide and/or carbon monoxide.

On its way in, Comet ISON played games with its predicted brightness curve, occasionally dipping below expectations and leading to revisions of its ultimate brightness at perihelion. A study conducted in February 2013 used 1,897 observations to create a light curve; it was found that had the comet maintained its original rate of brightening it would have reached magnitude −17 at perihelion. It seems that the comet's early brightness bursts were due to the vaporization of carbon dioxide and/or monoxide. Factoring in its observed rates of brightness change it is possible that the comet will reach around magnitude −6 at perihelion—still a truly Great Comet and a magnificent spectacle. Nevertheless, comet brightness predictions are one thing, but the fickle nature of comets can produce quite unexpected behaviors, and a brighter perihelion may still occur (Fig. 9.2).

Towards the end of March 2013 Comet ISON, shining at the 15th magnitude, ended its westward trek in southeastern Auriga and curved back towards the east.

Comet C/2012 S1. 21 Sep. 2012, 01h 15m (UT)
0.4-m reflector f/3 + CCD, exp. 5x100sec, 2"/pix., crop 200%
ISON-Kislovodsk observatory
Observers: V. Nevski, A. Novichonok

Fig. 9.1 Discovery image of Comet ISON. (Photo courtesy of Vitali Nevski and Artyom Novichonok)

The comet was imaged by the Hubble Space Telescope on April 10, and researchers in the Hubble Comet ISON Imaging Science Team noted that preliminary measurements from the images suggested that the comet's nucleus was no larger than 5–6.5 km across. This was remarkably small, considering the high level of activity that had already been observed in the comet. The Hubble team also produced a magnificent image of Comet ISON, based on five images obtained on April 30, showing the comet against a deep space background of stars and galaxies (Figs. 9.3 and 9.4).

On May 3 Comet ISON passed just half a degree north of dwarf planet Ceres (magnitude 7.8). By June the comet was an evening object embroiled in the twilight. Uninhibited by Earth's atmosphere, NASA's Spitzer telescope imaged it on June 13 at different infrared wavelengths that brought out detail in the Sun-swept tail, replenished by 54,000 tons of dust daily and the gases produced as carbon dioxide vaporized off the comet's nucleus at the rate of 1,000 tons per day (Figs. 9.5 and 9.6).

At the time of this writing, in early July, the comet was lost in the glare of the evening twilight, underwent conjunction with the Sun in mid-July and was expected

Fig. 9.2 A sequence of images of Comet ISON taken with the Gemini Multi-Object Spectrograph at Gemini North—February 4, March 4, April 3 and May 4, 2013 (2.5 arcmin field of view). (Photo courtesy of Gemini Observatory/AURA)

Fig. 9.3 Comet on April 10, 2013. At the time, it was still 4.15 au (621 million kilometers) from the Sun. (NASA)

to be observable once again in August 2013. After almost a year of anticipation since its discovery, the time came for amateur astronomers to seriously prepare for the arrival of Comet ISON, a newcomer to the inner Solar System that stood a chance of becoming one of history's brightest comets.

Fig. 9.4 Composite image of Comet ISON and the deep space background based on Hubble Space Telescope images taken in April 2013. (NASA/STScI)

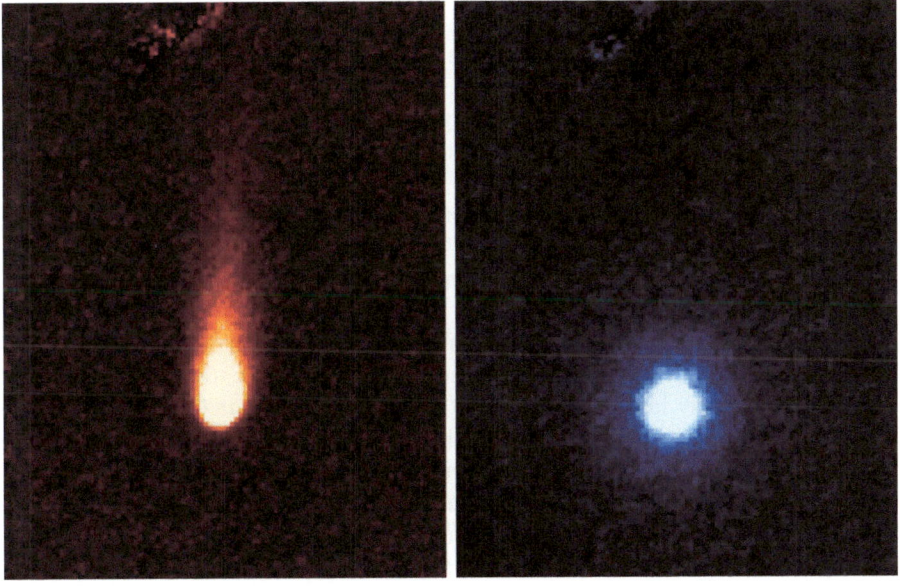

Fig. 9.5 Spitzer Space Telescope images of Comet ISON on June 13, 2003. (NASA)

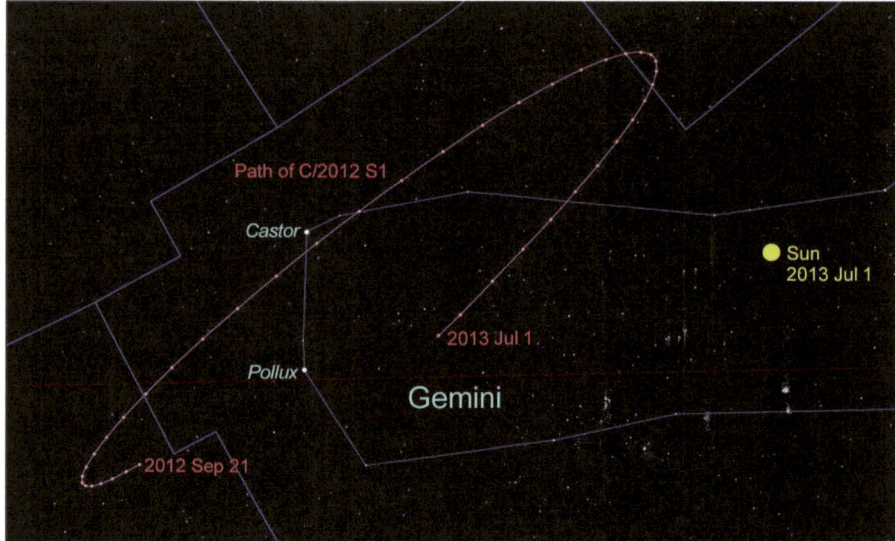

Fig. 9.6 The path of Comet ISON, marked at 7-day intervals, from its discovery on September 21, 2012, to July 1, 2013. The position of the Sun on July 1 is marked. (Illustration © by the author)

In August 2013 the comet was likely still to be relatively faint at around magnitude +12 and only become visible in the pre-dawn skies after solar conjunction in late August, when it would be in mid-Cancer. A viable target for both the astro-imager and the visual observer using a 6-in. telescope (or larger), this would be the starting point for amateur astronomers to begin their observing campaign.

On the morning of August 31, Comet ISON would lie a couple of degrees north of the bright open star cluster M44 (the Beehive Cluster) and be around 10° above the eastern horizon at the end of astronomical night. Delightful to view through big binoculars, it should also have been possible to see the northern part of M44 and the comet in the same field of view using an 8-in. SCT with a focal reducer and a 30-mm ultra-wide eyepiece.

The lead-up to possibly one of history's greatest cometary spectaculars offered exciting prospects. During September the comet would have traced a path that took it from mid-Cancer into western Leo. At the start of the month Comet ISON should have appeared 5½° east of Mars and shining at the 11th magnitude. The task of spotting the comet should have become easier as its angular distance from the Sun and its brightness increased, particularly during the first 3 weeks of the month when moonlight didn't present such a problem (the Moon was just 10° west of the comet on September 2, but should have appeared as a very dim and slender waning crescent, 26 days old). Mars would have steadily been closing in on the comet, and by September 27 be in conjunction with it, just 2° due south of the comet. By this

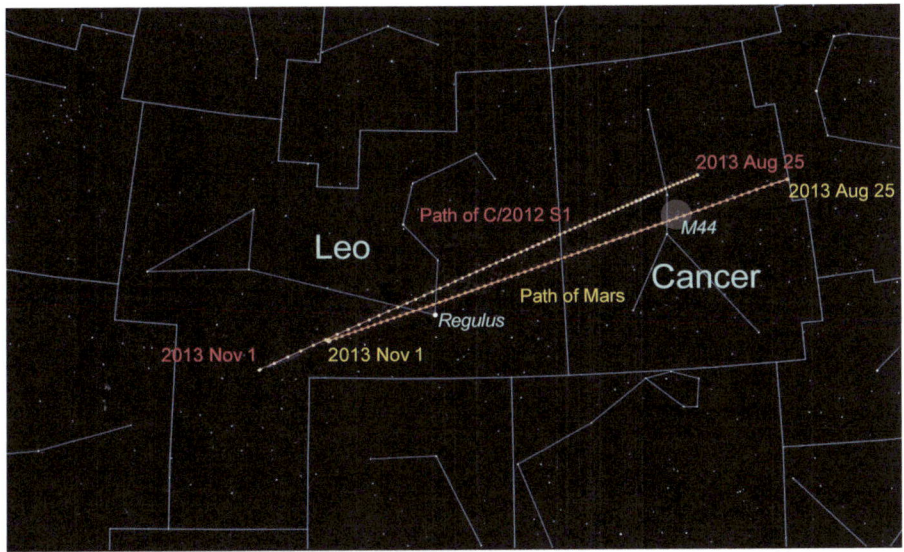

Fig. 9.7 The predicted paths of Comet ISON and Mars, marked at daily intervals, from August 25, 2013, to November 1, 2013. (Illustration © by the author)

time the comet likely shone at the 10th magnitude and could have been viewed with Mars in a single binocular field or suitably wide-angle telescopic field.

Another attraction in the vicinity should have been the galaxy NGC 2903 (magnitude +9.6), which lay 3½° north of the comet in Leo. Astro-imagers would have had fun capturing these objects at around this time (although the light of a waning crescent Moon would have interfered a little), some securing a single image showing Comet ISON flanked by NGC 2903 and Mars. At the end of September the comet had been projected to be less than 2° north of Mars (Fig. 9.7).

As well as being apparently close to each other, Comet ISON and Mars should have physically been in close proximity, at least on a cosmic scale. On September 1, Comet ISON was projected to be 0.7 au (105 million kilometers) from Mars and 3 au from Earth. By September 30, the comet would have closed its distance from Mars to just 0.08 au (11 million kilometers) and be 2.2 au from Earth. Comet ISON would have made its closest approach to Mars on October 2, when it was 0.07 au (10 million kilometers) from the Red Planet.

In mid-October the comet, now mostly likely at 10th magnitude, would have passed 2° north of Regulus in Leo. Two weeks later, at the beginning of November, it likely would have been an 8th magnitude object approaching Leo's southwestern border.

As it picked up speed, passing through southeastern Leo, across southern Virgo and Libra through November, Comet ISON would have brightened considerably. When it passed a degree north of galaxy NGC 4030 (magnitude 11.4) on November 9, the comet was likely shining at 7th magnitude. By mid-November, located south of

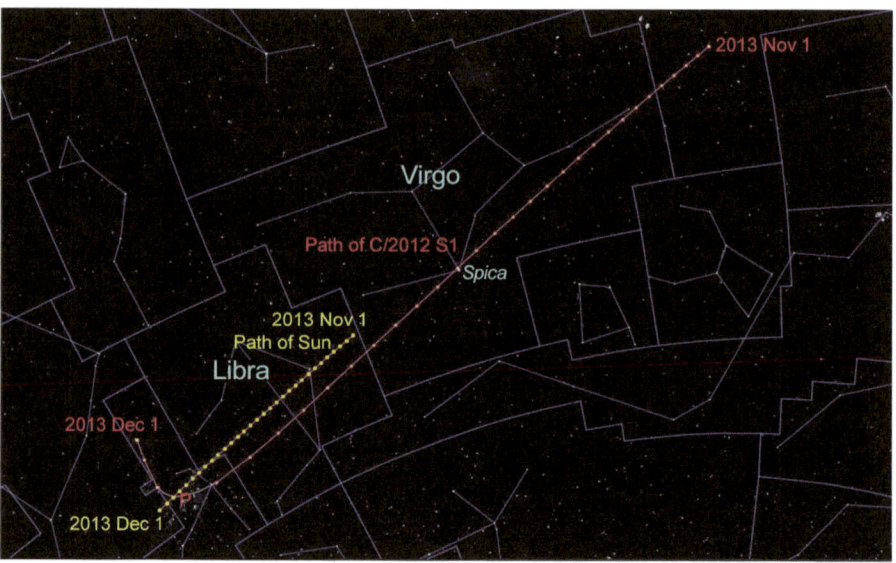

Fig. 9.8 The predicted paths of Comet ISON and the Sun, marked at daily intervals, from November 1 to December 1, 2013. P=perihelion. (Illustration © by the author)

the Bowl asterism of Virgo, it would have finally reached naked-eye visibility, and on November 17/18 it would have passed within just 17 arcmin of the star Spica.

Moving into western Libra on November 21, the comet by now was likely a bright 3rd magnitude object, displaying a lengthy tail and be a wonderful sight through binoculars. A morning object, the comet would now have been entering the dawn twilight, but its brightness would have kept it an easy object to locate. At zero magnitude on November 27, it would have entered minus magnitude levels as it rapidly curved around the Sun.

When the comet reached its perihelion of just 0.012 au (1.8 million kilometers) from the Sun's center on November 28, it might have reached anything from magnitude −6 to −13. Around a degree from the Sun, great care would have needed to be taken when observing it, since a direct view of the Sun through anything but properly protected optical aid could cause permanent eye damage or even blindness. The comet's tail, however, should have been easily visible rising above the horizon before sunrise prior to perihelion, and after sunset following perihelion (Fig. 9.8).

Rapidly rounding the Sun, Comet ISON and its north-projecting tail should have been seen in both the dawn and dusk skies between November 29 and mid-December, the tail protruding from the southeastern morning sky and also southwestern evening sky, emerging in the evening sky before the head of the comet itself had risen. Having established itself as an object favorable from the northern hemisphere, by mid-December the comet will likely be seen a good 44° north of the Sun, and, shining at 3rd magnitude, and might be viewed high in an astronomically dark morning sky. Moving north, the comet would have skirted the western border of Hercules,

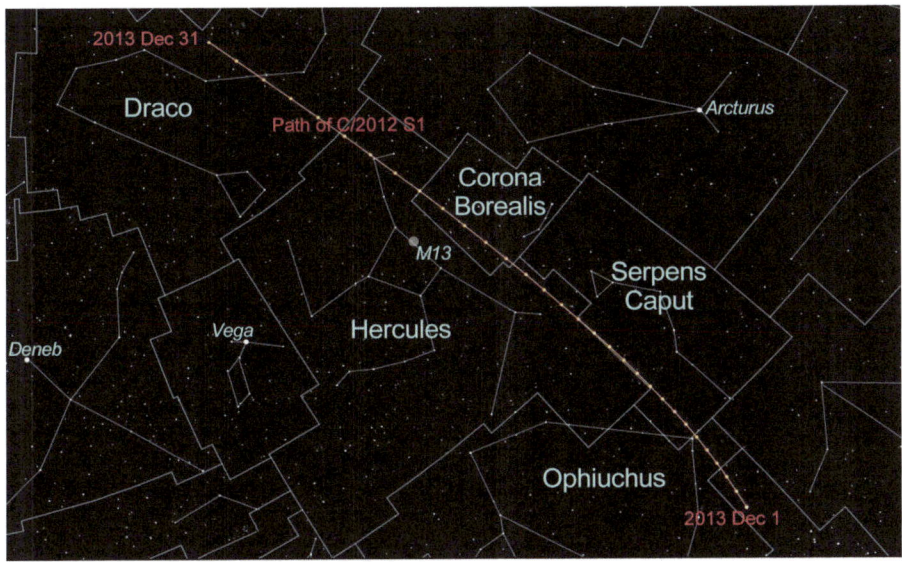

Fig. 9.9 The predicted path of Comet ISON, marked at daily intervals, from December 1 to December 31, 2013. (Illustration © by the author)

through the eastern part of Corona Borealis. The 4th magnitude comet should pass 5° west of the bright globular cluster M13 (magnitude 5.8) on December 22.

Towards the end of December the comet should become difficult to detect with the unaided eye from anywhere but a dark sky site. Late December will see the comet become a northern circumpolar object from northern temperate climes, observable all night. It will cross the border of Hercules into southern Draco on December 26, and by the year's end it will only be seen with the naked eye from a dark sky site, its tail (possibly 10° long and streaming through Ursa Minor) visible through binoculars (Fig. 9.9).

January 2, 2014, should see Comet ISON enter Ursa Minor, and by now it will probably be visible only through binoculars and telescopes, shining at magnitude 6. Within just 4° of the north celestial pole the comet should be just 2½° from Polaris on January 7 and on the following evening should pass open star cluster NGC 188 (magnitude 8.1). Briefly crossing the northeastern corner of Cassiopeia on January 13, Comet ISON, now 7th magnitude, will likely glide less than a degree west of galaxy IC 334 (magnitude 12.5) in neighboring Camelopardalis. It will then encounter another galaxy in Camelopardalis, IC 342 (magnitude 9.1) on January 19, this time passing just half a degree to its east. Encounters with two more deep sky objects in Camelopardalis should follow before the month is out—star cluster NGC 1502 (magnitude 6.9) on January 24 and NGC 361 (magnitude 11.7) on January 30. By now the comet should be shining around magnitude 9, and it would be interesting to compare the comet's brightness with these two compact clusters of different magnitudes.

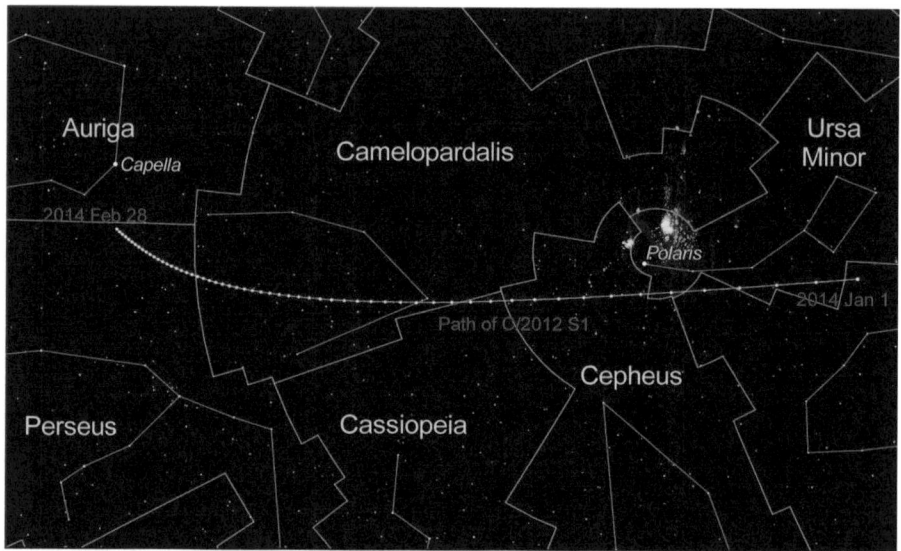

Fig. 9.10 The path of Comet ISON, marked at daily intervals, from January 1 to February 28, 2014. (Illustration © by the author)

As an aside, Earth should pass near the orbit of Comet ISON on January 14–15, 2014, some 3 months after the comet passes through that point in space. It is possible—although extremely unlikely—that our planet will encounter micron-sized dust particles left in the comet's wake, producing a meteor shower or a display of noctilucent clouds. But don't hold your breath!

Having entered Perseus on February 9, Comet ISON should track through a region rich in star clusters. NGC 1624 (magnitude 11.8) is the nearest one that the comet will approach, the pair being separated by 1° on February 16. By the end of February the comet should be near the northeastern border of Perseus and be an 11th magnitude object. Most amateurs interested in visual cometary observation will have probably wound down their Comet ISON observing program by this time, but the competent CCD imager should be able to squeeze more out of the comet until May, when it has crossed most of Auriga and shines at the 15th magnitude (Fig. 9.10).

C/2012K1 (PanSTARRS), 2014

Discovered on May 19, 2012, when it was 8.7 au from the Sun, C/2012K1 should be an object favorable from the northern hemisphere through its perihelion in August 2014, in mid-September crossing the celestial equator to become a comet best viewed from the southern hemisphere.

Having brightened to the 9th magnitude by April 1, 2014, the object should be a good subject for amateur telescopic scrutiny. From the northwestern corner of

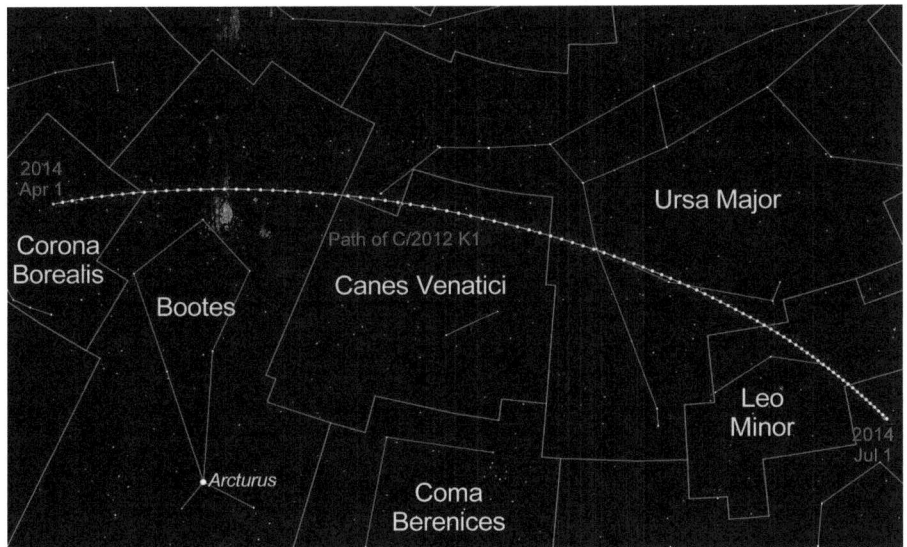

Fig. 9.11 The predicted path of C/2012K1 (PanSTARRS) marked at daily intervals, from April 1 to July 1, 2014. (Illustration © by the author)

Corona Borealis the comet should move into Boötes on April 10, passing 3° north of the tip of the Kite asterism on the 16th. By then it should be a circumpolar object for northern temperate observers, at its highest in the sky around 2.30 local time. On May 1 the comet, shining at around magnitude 8.5, should pass just 2° north of M51 (magnitude 8), the famous Whirlpool Galaxy in Canes Venatici. On May 14 the comet should cross the border into Ursa Major, and over the course of the ensuing weeks pass through a field rich in faint galaxies, notably passing a few arcminutes by NGC 3949 (magnitude 11.4) on May 16/17, NGC 3726 (magnitude 10.9) on May 20, NGC 3614 (magnitude 15), which likely will be enveloped by the coma on May 23/24 and NGC 3319 (magnitude 11.5) on June 2/3.

On June 7 C/2012K1 should enter northern Leo Minor and shine at magnitude 7.5 as it passes extremely close to the star 21 Leonis Minoris (magnitude 4.5) on the evening of June 15. Despite the very short nights of northern midsummer, the comet should still be visible in a dark sky at a respectable altitude above the northwestern horizon after sunset; however, twilight and low altitude might make the comet difficult to observe once it has entered northwestern Leo on June 22 (Fig. 9.11).

A perihelion of 1.05 au should be reached on August 27, 2014, and observations can recommence in the morning in mid-September when it should be in western Hydra and shining at magnitude 6. It should be able to be seen from northern climes, even through to its peak brightness in mid-October, when it will just be within naked-eye visibility at magnitude 5.9; it should then be located just half a degree east of the bright open star cluster NGC 2467 (magnitude 7.1).

The comet should at this time be increasingly more favorable to view from the southern hemisphere as it passes through Puppis, passing half a degree west of another

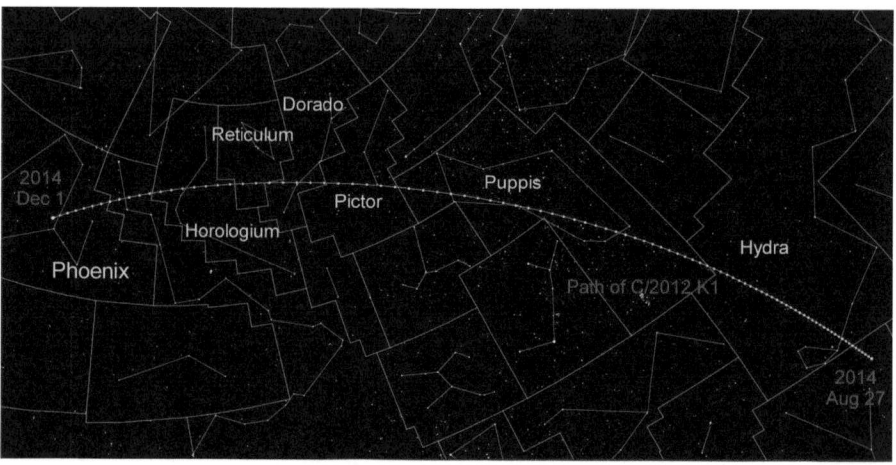

Fig. 9.12 The predicted path of C/2012K1 (PanSTARRS) marked at daily intervals, from August 27 (perihelion) to December 1, 2014. (Illustration © by the author)

bright open star cluster, NGC 2499 (magnitude 6) on October 19. Having faded from naked eye view, on November 1 it should enter northeastern Pictor, and a week later be immersed in a galaxy-rich part of the sky, passing such fine examples as NGC 1566 (magnitude 10.2), NGC 1549 (magnitude 10.6) and NGC 1553 (magnitude 10) on November 9–10. After crossing Reticulum and into Horologium on November 13, it should pass within a degree of the globular cluster NGC 1261 (magnitude 8.3).

By November 6, C/2012K1 should shine at the 7th magnitude and pass galaxy NGC 1705 (magnitude 12.8) by 1°. From some southerly locations in Australia, New Zealand and South America the comet should be a circumpolar object for much of November. When in Phoenix in early December the comet will likely fall below magnitude 8. It should continue to head westward but be gaining in declination, winding down its apparition in Sculptor and reaching 12th magnitude by the end of January 2015 (Fig. 9.12).

Periodic Comets

Hundreds of comets have well-known orbits lying within the realm of the planets, and they make regular, predictable visits to the inner Solar System. Although most of these are too faint to observe through average backyard telescopes even when they are at their brightest near perihelion, a number of them are familiar to amateur astronomers. Periodic comet Halley, for example, has a 76-year orbital period and aphelion of some 35 au (further than the orbit of Neptune). Orbiting the Sun every 3.3 years, periodic comet Encke has the shortest known period of any comet, its orbital path taking it from aphelion inside the orbit of Jupiter to perihelion inside Mercury's orbit.

Named after their discoverer (or the discoverer of their periodicity), a total of 291 periodic comets are currently designated a number on a list maintained by the Minor Planet Center and prefixed with P (for periodic), in order of the discovery of their periodic nature. Thus, Halley's Comet is referred to as 1P/Halley, Encke's Comet 2/P Encke and so on. In addition, there are more than 160 periodic comets without a designated number and an additional several dozen asteroids within the Solar System that display all the attributes of cometary nuclei.

2P/Encke (Encke's Comet), 2013, 2017

This famous comet with a period of just 3.3 years came into our vicinity in late 2013, having approached Earth at its closest on October 17, 2013 (shining at just 9th magnitude) and coming to perihelion (0.34 au) on November 21, 2013. A full 18° west of the Sun at this time, it was likely at its brightest, a naked-eye object of the 5th magnitude, in the morning skies in western Libra, and, by a stroke of cosmic coincidence, just 7° east of C/2012S1 (ISON). The two comets would have appeared to move closer to each other on subsequent mornings, reaching their minimum separation on the morning of November 24, when they were likely less than 2° apart. Although the observing circumstances weren't necessarily ideal, with the presence of morning twilight and a waning Moon high in the southwest, this posed a rare opportunity to see two comets in the same low power field of view. Of course, shining at the 5th magnitude Encke's Comet may have proved difficult to perceive, but C/2012S1 should have been a bright 3rd magnitude object (Fig. 9.13).

The next apparition of Encke's Comet should take place in 2017. Initially an evening object it will likely curve through western Pisces during February, brightening from the 11th to the 6th magnitude over the course of the month, its angular distance from the Sun decreasing from 46 to 25° in the same period; thankfully, moonlight shouldn't present a problem for most of February, but twilight will likely become more troublesome towards the month's end. Perihelion should take place on March 10, 2017 (0.34 au), with perigee on the following day, but the comet will likely be unobservable at this time due to its proximity to the Sun. It might be picked up from the southern hemisphere as a morning object of the 6th magnitude in late March as it heads through western Aquarius, fading to the 11th magnitude by early April as its separation from the Sun increases (Fig. 9.14).

154P/Brewington, 2013–2014

154P/Brewington is an evening object, visible in both northern and southern hemispheres, that will trace a northwestward path from northern Aquarius to southern Pegasus between November 1, 2013, and January 1, 2014. Shining at 10th magnitude in early November, it should brighten to magnitude 9 at perihelion (1.6 au) on

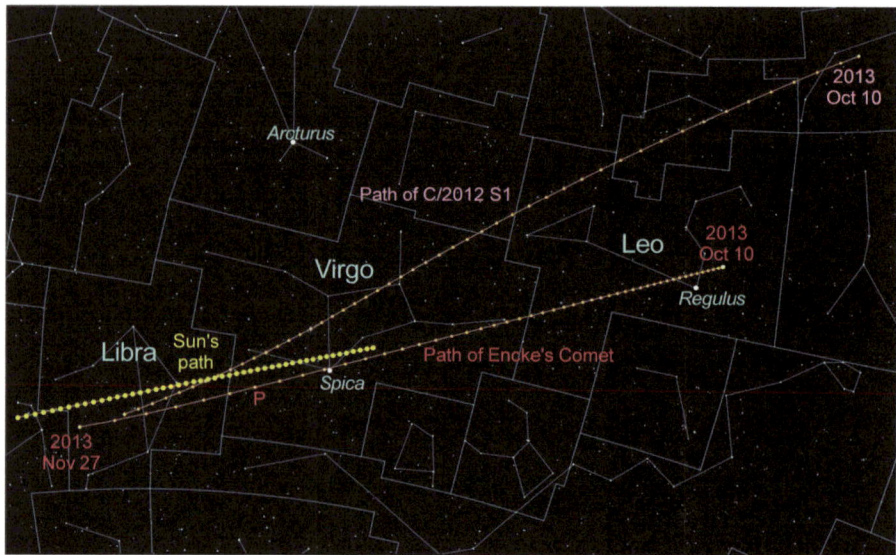

Fig. 9.13 The predicted paths of 2P/Encke, C/2012S1 (ISON) and the Sun, marked in daily increments, between October 10 and November 27, 2013. P=perihelion. (Illustration © by the author)

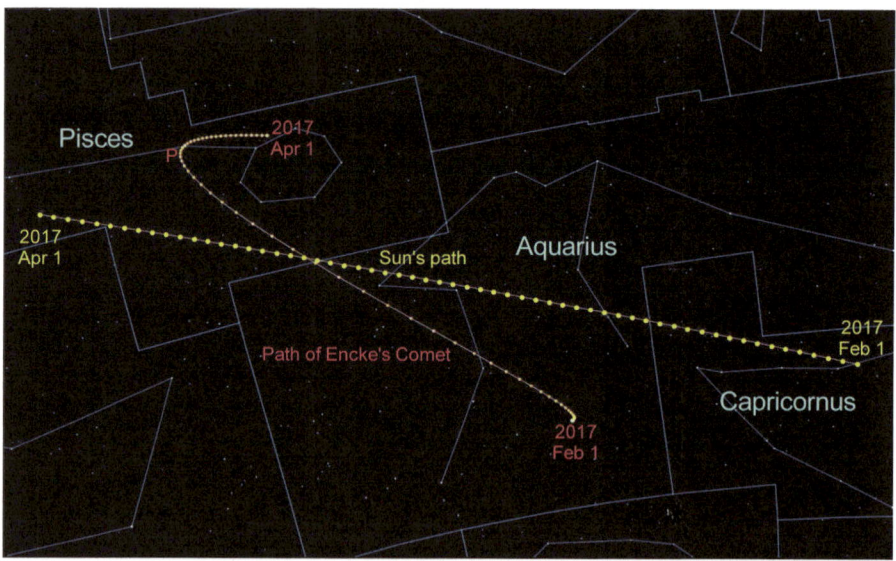

Fig. 9.14 The predicted paths of 2P/Encke and the Sun, marked in daily increments, between February 1 and April 1, 2017. P=perihelion. (Illustration © by the author)

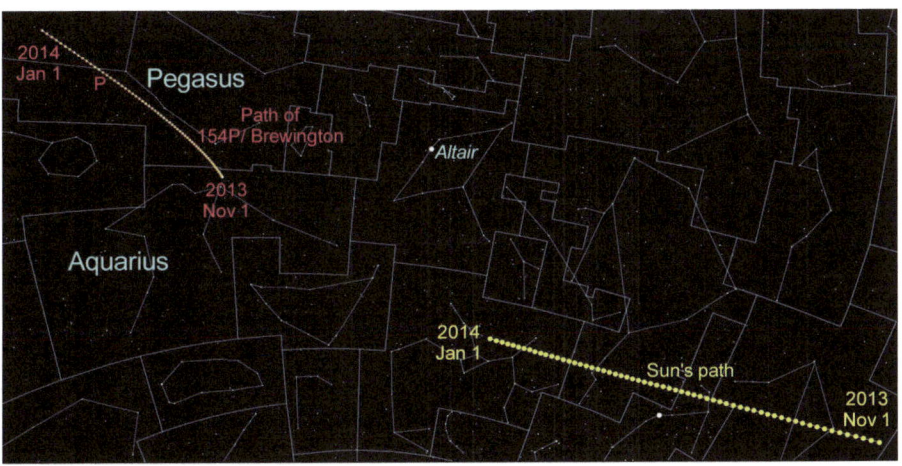

Fig. 9.15 The paths of 154P/Brewington and the Sun, marked in daily increments, between November 1, 2013, and January 1, 2014. P=Perihelion. (Illustration © by the author)

December 12, when it should be found just 2° south of Markab (Alpha Pegasi), the bottom right hand star in the Square of Pegasus asterism.

By January 1, 2014, the comet should have fallen slightly in brightness to the 10th magnitude and its angular separation from the Sun would be 58°. It should remain at a respectable distance from the Sun, unconquered by evening twilight, for quite a considerable time. On February 9, 145P/Brewington, likely shining at 11th magnitude, should pass just 1° south of the famous Triangulum Galaxy, M33 (magnitude 6.2), an excellent viewing and imaging opportunity with both objects able to be encompassed in a single binocular or rich telescopic field of view for a day or two either side of this date (Fig. 9.15).

290P/Jäger

Comet 290P/Jäger is excellently placed for observation from the northern hemisphere at this apparition as it heads south through Auriga and Gemini. On February 1, 2014, it should be found in southeastern Auriga, a 10th magnitude object shining at its peak brightness, and between February 21 and 26 it should pass within 3° northeast of the bright open cluster M35 (magnitude 5.5) in Gemini. Perihelion will take place on March 12, when it is in western Gemini, 2½° west of Jupiter. Shining at 11th magnitude, it should proceed to curve southwest of Jupiter, passing the giant planet within just 2° between March 16 and 20. By April 1 Comet Jäger will likely be less than 2° northeast of open cluster NGC 2304 (magnitude 10) and less than 2° west of the galaxy designated NGC 2339 (magnitude 12.5), and it should pass

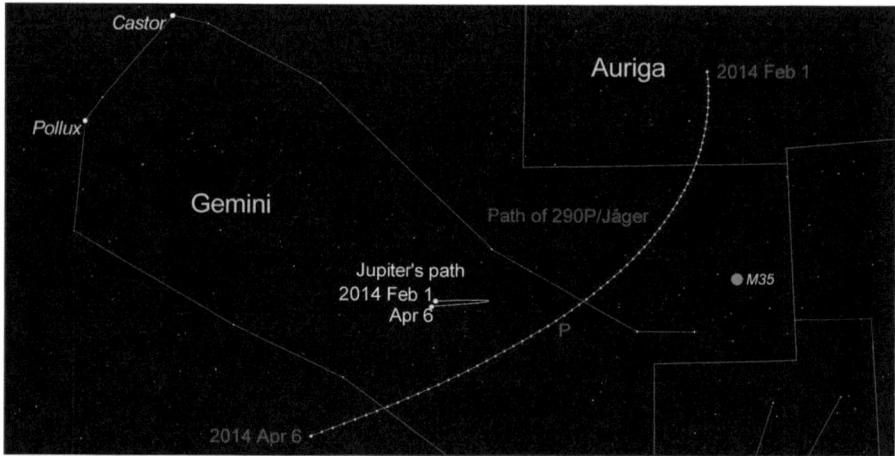

Fig. 9.16 The predicted paths of 290P/Jäger and Jupiter, between February 1 and April 6, 2014. P=perihelion. (Illustration © by the author)

just 14 arcmin south of the galaxy on April 6. (The latter will be challenging to observe because of the light of the nearby first quarter Moon.) (Fig. 9.16).

4P/Faye, 2014

With a period of 7.5 years, the 2014 apparition of Comet Faye should be far from spectacular. Indeed, its likely maximum brightness of magnitude 12.5 at perihelion on May 29, 2014, makes it the dimmest in this list. However, it's mentioned because this comet was the first periodic comet ever to have been named after its discoverer, Hervé Faye at the Royal Observatory in Paris, back in November 1843. Just 15° west of the Sun this perihelion, it's unlikely to be visible in the morning twilight, and even though it distances itself from the Sun in the ensuing days and weeks, its brightness will be continually falling. Best seen from the southern hemisphere, perhaps the most photogenic opportunity to view it—given dark skies and a good instrument—should be on the morning of June 25, when it passes very near open star cluster NGC 1647 in Taurus, with a very old waning crescent Moon just 5° to the east (Fig. 9.17).

15P/Finlay, 2014–15

An evening object during this apparition, best viewed from southern climes. 60° east of the Sun and likely visible as a 12th magnitude object in mid-Capricornus in early December 2014, Comet 15 P/Finlay heads east, likely brightening to magnitude 11 by perihelion (0.97 au) in eastern Capricornus on December 27. It should

Fig. 9.17 Simulation of Comet Faye near open star cluster NGC 1647 on the morning of June 25, 2014, as seen from the southern hemisphere (1° field of view). (Illustration © by the author)

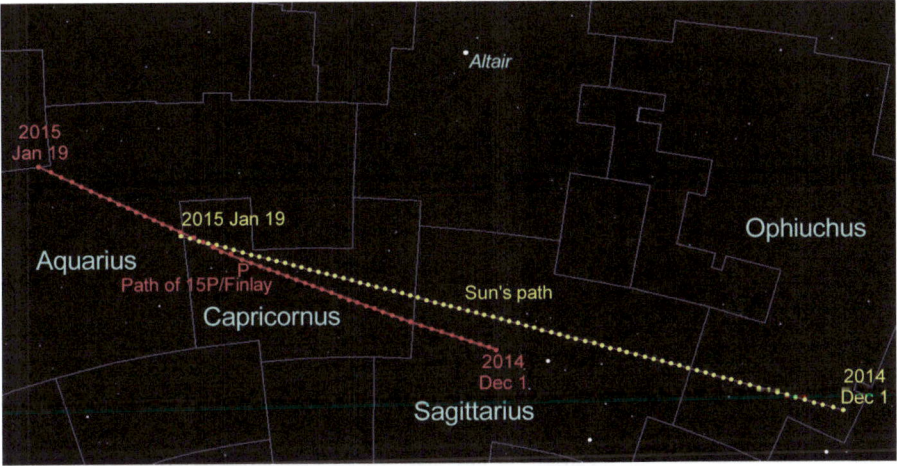

Fig. 9.18 The paths of 15P/Finlay and the Sun, marked in daily increments, between July 1, 2014, and September 1, 2016. P=perihelion. (Illustration © by the author)

proceed through Aquarius and be at its brightest at magnitude 10 by late January 2015, when it will be located south of the Circlet asterism in western Pisces, 45° east of the Sun. 15P/Finlay will have faded back to the 12th magnitude by early March when it is in Aries (Fig. 9.18).

10P/Tempel (Tempel 2), 2015

With a period 5.4 years, this comet's next apparition takes place in late 2015. Although it will be exclusively an object for study with large binoculars and telescopes, its sizable separation from the Sun and high altitude in the evening skies of the southern hemisphere on the approach towards perihelion for a month or so and following perihelion makes it worth taking a look at. From October 1, 2015, to perihelion on November 14, the comet will likely head from southern Ophiuchus into mid-Sagittarius, remaining at the 10th magnitude and brightening just half a magnitude. At perihelion it should be a good 50° east of the Sun. 10P/Tempel can be followed until late December as it moves into Capricornus, but the evening twilight then begins to interferes (Fig. 9.19).

9P/Tempel (Tempel 1), 2016

Favorable and unfavorable returns of 9P/Tempel take place on an alternating basis through 2022. The comet's next favorable return takes place in 2016, with its closest approach to Earth (0.86 au) on April 21, when it shines at the 12th magnitude in eastern Leo. It should slowly brighten, and at a perihelion (1.51 au) in mid-Virgo on July 21, 2016, it should be a 10th magnitude object (Fig. 9.20).

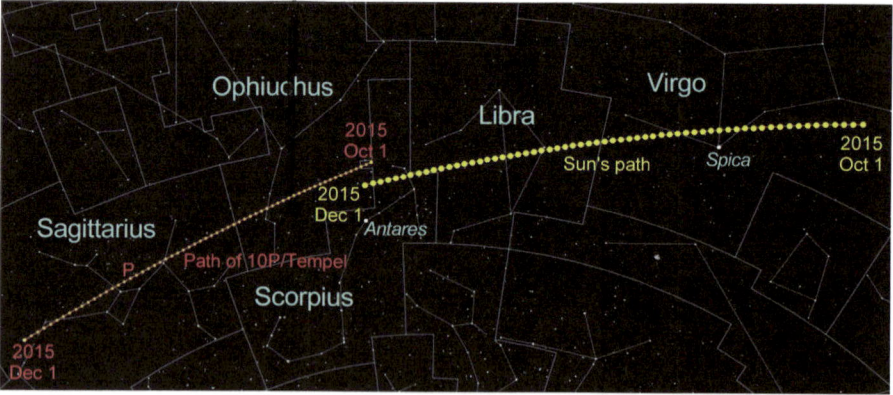

Fig. 9.19 The predicted paths of 10P/Tempel and the Sun, marked in daily increments, between October 1 and December 1, 2015. P=perihelion. (Illustration © by the author)

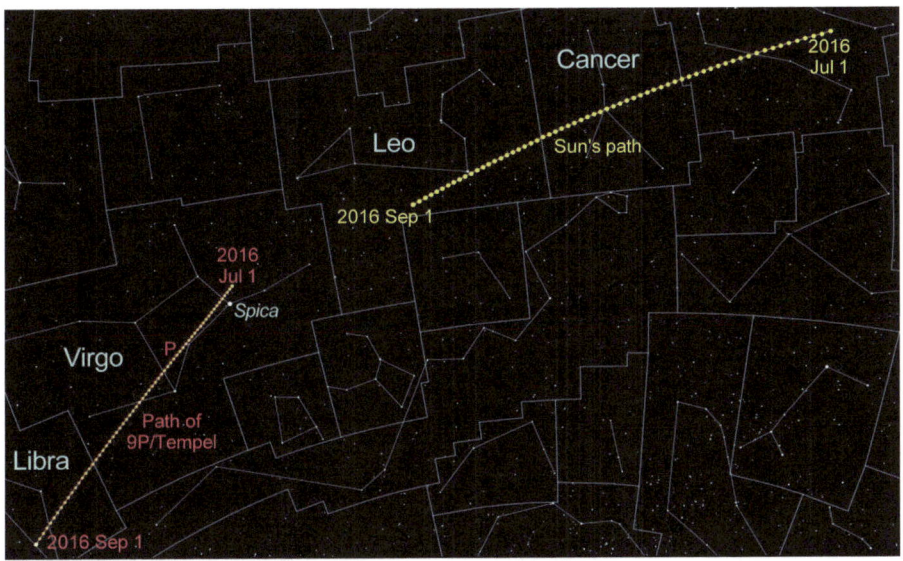

Fig. 9.20 The paths of 9P/Tempel and the Sun, marked in daily increments, between July 1 and September 1, 2016. P=perihelion. (Illustration © by the author)

An object for large binocular and telescopic studies, it won't likely be particularly bright, but a number of things make its worth mentioning. At perihelion it should be a good 80° east of the Sun, which means that it can be seen high in an astronomically dark sky well into the night, and its location near the celestial equator means that it should be seen well from both hemispheres.

This comet is also famous for having been visited by NASA's Deep Impact (2005) and Stardust-NEXT (2011) space probes. Furthermore, 9P/Tempel will likely undergo an orbital change. On May 26, 2024, the comet should pass 0.55 au from Jupiter, the gravitational effects of which should increase the perihelion distance to 1.77 au, along with a lengthening of its orbital period to 6 years. This means that 9P/Tempel will likely undergo unfavorable returns in 2028 and 2034. After another close approach to Jupiter on April 17, 2036 (0.91 au) the comet's perihelion distance should increase further to 1.93 au, giving it a period of 6.3 years. The larger perihelion distance means that 9P/Tempel's subsequent apparitions won't be nearly so bright.

Table 9.1 Numbered periodic comets

Designation (informal/other name)	Discovered	Period (yrs)	Next perihelion	Notes
1P/Halley (Halley's Comet)		*75.3*	*2061 Jul 28*	*Visit*
2P/Encke (Encke's Comet)	*1786 Jan 17*	*3.3*	*2013 Nov 21*	
3D/Biela	*1772 Mar 08*			*Dead*
4P/Faye	**1843 Nov 23**	**7.5**	**2014 May 29**	
5D/Brorsen	*1846 Feb 26*			Lost
6P/d'Arrest	1851 Jun 28	6.5	2015 Mar 2	
7P/Pons-Winnecke	1819 Jun 12	6.4	2015 Jan 30	
8P/Tuttle	1790 Jan 09	13.6	2021 Aug 27	
9P/Tempel (Tempel 1)	**1867 Apr 03**	**5.6**	**2016 Jul 21**	**Visit**
10P/Tempel (Tempel 2)	**1873 Jul 03**	**5.4**	**2015 Nov 14**	
11P/Tempel-Swift-LINEAR	1869 Nov 27	6.4	2014 Aug 26	
12P/Pons-Brooks	1812 Jul 21	70.9	2024 Apr 21	
13P/Olbers	*1815 Mar 06*	*72–77*	*2024 Jun 30*	
14P/Wolf	1884 Sep 17	8.7	2017 Dec 01	
15P/Finlay	**1886 Sep 26**	**6.8**	**2014 Dec 27**	
16P/Brooks (Brooks 2)	1889 Jul 07	6.1	2014 Jun 7	
17P/Holmes	*1892 Nov 07*	*6.9*	*2014 Mar 27*	
18D/Perrine-Mrkos	1896 Dec 09			Lost
19P/Borrelly	1904 Dec 28	6.8	2015 May 28	Visit
20D/Westphal	1852 Jul 24			Lost
21P/Giacobini-Zinner	*1900 Dec 20*	*6.6*	*2018 Sep 10*	*Visit*
22P/Kopff	1906 Aug 22	6.5	2015 Oct 25	
23P/Brorsen-Metcalf	*1847 Jul 20*	*70.5*	*2059 Jun 08*	
24P/Schaumasse	1911 Dec 01	8.2	2017 Nov 16	
25D/Neujmin (Neujmin 2)	1916 Feb 24			Lost
26P/Grigg-Skjellerup	1902 Jul 23	5.3	2013 Jul 06	Visited
27P/Crommelin	1818 Feb 23	27.4	2039 May 27	
28P/Neujmin (Neujmin 1)	1913 Sep 04	18.2	2021 Mar 11	
29P/Schwassmann-Wachmann	1927 Nov 15	14.7	2019 Mar 07	
30P/Reinmuth 1 (Reinmuth 1)	1928 Feb 22	7.3	2017 Aug 19	
31P/Schwassmann-Wachmann	1929 Jan 17	8.7	2019 Jul 06	
32P/Comas Solà	1926 Nov 06	8.8	2014 Oct 17	
33P/Daniel	1909 Dec 07	8.1	2016 Aug 22	
34D/Gale	1927 Jun 07			Lost
35P/Herschel-Rigollet	*1788 Dec 21*	*155*	*2092 Feb 17*	
36P/Whipple	1933 Oct 15	8.5	2020 May 31	
37P/Forbes	1929 Aug 01	6.3	2018 May 03	
38P/Stephan-Oterma	1867 Jan 22	37.7	2018 Nov 10	
39P/Oterma	1943 Apr 08	19.4	2023 Jul 11	
40P/Vaisala (Vaisala 1)	1939 Feb 08	10.4	2014 Nov 15	
41P/Tuttle-Giacobini-Kresák	1858 May 03	5.4	2017 Apr 13	
42P/Neujmin (Neujmin 3)	1929 Aug 02	10.7	2015 Apr 8	
43P/Wolf-Harrington	1924 Dec 22	6.1	2016 Aug 19	

(continued)

Table 9.1 (continued)

Designation (informal/other name)	Discovered	Period (yrs)	Next perihelion	Notes
44P/Reinmuth (Reinmuth 2)	1947 Sep 10	10.9	2014 Nov 15	
45P/Honda-Mrkos-Pajdušáková	1948 Dec 03	5.3	2016 Dec 31	
46P/Wirtanen	1948 Jan 15	5.4	2018 Dec 12	
47P/Ashbrook-Jackson	1948 Aug 26	8.3	2017 Jun 10	
48P/Johnson	1949 Aug 25	7.0	2018 Aug 12	
49P/Arend-Rigaux	1951 Feb 05	6.7	2018 Jul 15	
50P/Arend	1951 Oct 04	8.3	2016 Feb 08	
51P/Harrington	1953 Aug 14	7.0	2015 Aug 12	
52P/Harrington-Abell	1955 Mar 22	7.4	2014 Mar 07	
53P/Van Biesbroeck	1954 Sep 01	12.5	2016 Apr 29	
54P/de Vico-Swift-NEAT	1844 Aug 23	7.3	2017 Apr 15	
55P/Tempel-Tuttle	*1866 Jan 06*	*33.2*	*2031 May 20*	
56P/Slaughter-Burnham	1959 Jan 27	11.5	2016 Jul 18	
57P/duToit-Neujmin-Delporte	1941 Jul 18	6.4	2015 May 22	
58P/Jackson-Neujmin	1936 Sep 20	8.1	2020 May 25	
59P/Kearns-Kwee	1963 Aug 17	9.5	2018 Sep 16	
60P/Tsuchinshan (Tsuchinshan 2)	1965 Jan 11	6.8	2018 Dec 11	
61P/Shajn-Schaldach	1949 Sep 18	7.1	2015 Oct 02	
62P/Tsuchinshan 1 (Tsuchinshan 1)	1965 Jan 01	6.6	2017 Nov 16	
63P/Wild (Wild 1)	1960 Mar 26	13.2	2026 Jul 06	
64P/Swift-Gehrels	1889 Nov 16	9.3	2018 Nov 03	
65P/Gunn	1970 Oct 27	6.8	2017 Oct 16	
66P/du Toit	1944 May 16	14.7	2018 May 19	
67P/Churyumov-Gerasimenko	1969 Oct 22	6.5	2015 Aug 13	
68P/Klemola	1965 Nov 15	10.8	2019 Nov 09	
69P/Taylor	1915 Nov 24	7.0	2019 Mar 18	
70P/Kojima	1970 Dec 27	7.1	2014 Oct 20	
71P/Clark	1973 Jun 09	5.5	2017 Jun 30	
72D/Denning-Fujikawa	1881 Oct 04			Lost
73P/Schwassmann-Wachmann	1930 May 02	5.4	2017 Mar 16	
74P/Smirnova-Chernykh	1975 Mar 04	8.5	2018 Jan 26	
75D/Kohoutek	1975 Feb			Lost
76P/West-Kohoutek-Ikemura	1975 Jan	6.5	2019 Oct 26	
77P/Longmore	1975 Jun 10	6.8	2016 May 13	
78P/Gehrels (Gehrels 2)	1973 Sep 29	7.2	2019 Apr 02	
79P/du Toit-Hartley	1945 Apr 09	5.1	2018 Sep 13	
80P/Peters-Hartley	1846 Jun 26	8.1	2014 Nov 10	
81P/Wild (Wild 2)	1978 Jan 06	6.4	2016 Jul 20	Visited
82P/Gehrels (Gehrels 3)	1975 Oct 27	8.4	2018 Jun 28	
83D/Russell (Russell 1)	1979 Jun 16			Lost
84P/Giclas	1978 Sep 08	7.0	2020 Jun 03	
85P/Boethin	1975 Jan 04	5.6	2020 Jul 29	?Lost
86P/Wild (Wild 3)	1980 May	6.9	2015 Apr 03	

(continued)

Table 9.1 (continued)

Designation (informal/other name)	Discovered	Period (yrs)	Next perihelion	Notes
87P/Bus	1981 Mar 02	6.5	2013 Dec 19	
88P/Howell	1981 Aug 29	5.5	2015 Apr 06	
89P/Russell (Russell 2)	1980 Sep 28	7.4	2016 Dec 14	
90P/Gehrels (Gehrels 1)	1972 Oct 11	14.8	2017 Jun 19	
91P/Russell (Russell 3)	1983 Jun 14	7.7	2020 Nov 09	
92P/Sanguin	1977 Oct 15	12.4	2015 Mar 01	
93P/Lovas (Lovas 1)	1980 Dec 05	9.2	2017 Mar 01	
94P/Russell (Russell 4)	1984 Mar 07	6.6	2016 Oct 27	
95P/Chiron / 2060 Chiron	1977 Oct 18	50.7	2046 Aug 03	
96P/Machholz (Machholz 1)	*1986 May 12*	*5.3*	*2017 Oct 27*	
97P/Metcalf-Brewington	1906 Nov 15	10.5	2022 Feb 15	
98P/Takamizawa	1984 Jul 30	7.4	2021 Jan 04	
99P/Kowal (Kowal 1)	1977 Apr 24	15.1	2022 Apr 12	
100P/Hartley (Hartley 1)	1985 Jun 13	6.3	2016 Apr 02	
101P/Chernykh	1977 Aug 19	13.9	2020 Jan 13	
102P/Shoemaker (Shoemaker 1)	1984 Sep 27	7.2	2021 Jan 22	
103P/Hartley (Hartley 2)	1986 Mar 15	6.5	2017 Apr 20	Visited
104P/Kowal (Kowal 2)	1979 Jan 27	6.2	2016 Mar 26	
105P/Singer Brewster	1986 May 03	6.5	2018 Aug 10	
106P/Schuster	1975 Oct 09	7.3	2014 Jul 20	
107P/Wilson–Harrington	1949 Nov 19	4.3	2014 Feb 05	
108P/Ciffréo	1985 Nov 08	7.3	2014 Oct 18	
109P/Swift-Tuttle	*1862 Jul 16*	*133.3*	*2126 Jul 12*	
110P/Hartley (Hartley 3)	1988 Feb 19	6.9	2014 Dec 17	
111P/Helin-Roman-Crockett	1989 Jan 05	8.1	2021 Jun 15	
112P/Urata-Niijima	1986 Nov 03	6.7	2020 Feb 07	
113P/Spitaler	1890 Nov 17	7.1	2015 Apr 23	
114P/Wiseman-Skiff	1987 January	6.7	2020 Jan 14	
115P/Maury	1985 Aug 16	8.8	2020 Jul 29	
116P/Wild (Wild 4)	1990 Jan 21	6.5	2016 Jan 11	
117P/Helin-Roman-Alu (H-R-A 1)	1989 Oct 02	8.2	2014 Mar 27	
118P/Shoemaker-Levy (S-L 4)	1991 Feb 09	6.4	2016 Jun 16	
119P/Parker-Hartley	1989 Mar 02	8.9	2014 Apr 02	
120P/Mueller (Mueller 1)	1987 Oct 18	8.4	2021 May 07	
121P/Shoemaker-Holt (S-H 2)	1989 Mar 09	9.9	2023 Jun 28	
122P/de Vico	1846 Feb 20	74.4	2069 Oct 21	
123P/West-Hartley	1989 May 11	7.6	2019 Feb 05	
124P/Mrkos	1991 Mar 16	6.0	2014 Apr 09	
125P/Spacewatch	1991 Sep 08	5.5	2018 Aug 27	
126P/IRAS	1983 Jun 28	13.4	2023 Jul 05	
127P/Holt-Olmstead	1990 Sep 14	6.4	2016 Mar 17	
128P/Shoemaker-Holt (S-H 1)	1987 Oct 18	9.6	2017 Jan 10	
129P/Shoemaker-Levy (S-L 3)	1991 Feb 7	8.9	2014 Feb 11	

(continued)

Table 9.1 (continued)

Designation (informal/other name)	Discovered	Period (yrs)	Next perihelion	Notes
130P/McNaught-Hughes	1991 Sep 30	6.7	2018 Jan 21	
131P/Mueller (Mueller 2)	1990 Sep 15	7.1	2019 Jan 24	
132P/Helin-Roman-Alu (H-R-A 2)	1989 Oct 26	8.3	2014 May 21	
133P/Elst-Pizarro	1996 Jul 14	5.6	2018 Sep 20	
134P/Kowal-Vávrová	1983 May 08	15.6	2014 May 21	
135P/Shoemaker-Levy (S-L 8)	1992 Apr 05	7.5	2014 Nov 01	
136P/Mueller (Mueller 3)	1990 Sep 24	8.6	2016 May 31	
137P/Shoemaker-Levy (S-L 2)	1990 Oct 25	9.6	2018 Dec 13	
138P/Shoemaker-Levy (S-L 7)	1991 Nov 13	6.9	2019 May 02	
139P/Väisälä-Oterma	1939 Oct 07	9.6	2017 Dec 10	
140P/Bowell-Skiff	1983 Feb 11	16.2	2015 Aug 08	
141P/Machholz (Machholz 2)	1994 Aug 13	5.2	2015 Aug 24	
142P/Ge-Wang	1988 Nov 04	11.1	2021 May 12	
143P/Kowal-Mrkos	1984 Apr 23	8.9	2018 May 07	
144P/Kushida	1994 Jan 08	7.6	2016 Aug 30	
145P/Shoemaker-Levy (S-L 5)	1993 Dec 08	8.4	2017 Aug 31	
146P/Shoemaker-LINEAR	1984 Sep 26	8.1	2016 Jun 30	
147P/Kushida-Muramatsu	1993 Dec 08	7.4	2016 Feb 27	
148P/Anderson-LINEAR	1963 Nov 22	7.1	2015 Jun 13	
149P/Mueller (Mueller 4)	1992 Apr 09	9.0	2019 Feb 16	
150P/LONEOS	2000 Nov 25	7.7	2016 Jul 24	
151P/Helin	1987 Aug 24	14.1	2015 Oct 08	
152P/Helin-Lawrence	1993 May 17	9.5	2022 Jan 13	
153P/Ikeya-Zhang	*2002 Feb 01*	*366.5*	*2362 Sep 01*	
154P/Brewington	**1992 Aug 28**	**10.7**	**2013 Dec 12**	
155P/Shoemaker (Shoemaker 3)	1986 Jan 10	17.1	2019 Nov 15	
156P/Russell-LINEAR	1986 Sep 03	6.8	2014 Apr 16	
157P/Tritton	1978 Feb 11	6.3	2016 Jun 10	
158P/Kowal-LINEAR	1979 Jul 24	10.3	2021 May 10	
159P/LONEOS	2003 Oct 16	14.3	2018 May 22	
160P/LINEAR	2004 Jul 15	8.0	2019 Dec 02	
161P/Hartley-IRAS	1983 Nov 04	21.5	2026 Nov 27	
162P/Siding Spring	2004 Nov 10	5.3	2015 Jul 11	
163P/NEAT	2004 Nov 05	7.3	2019 Aug 05	
164P/Christensen	2004 Dec 21	7.0	2018 May 31	
165P/LINEAR	2000 Jan 29	76.4	2075 Oct 14	
166P/NEAT	2001 Oct 15	51.4	2053 Nov 26	
167P/CINEOS	2004 Aug 10	65.2	2066 Mar 18	
168P/Hergenrother (Hergenrother 1)	1998 Nov 22	6.9	2019 Aug 05	
169P/NEAT	202 Mar 15	4.2	2014 Feb 15	
170P/Christensen	2005 Jun 17	8.6	2014 Sep 19	
171P/Spahr	1998 Nov 16	6.7	2019 Jan 13	
172P/Yeung	2002 Jan 21	6.6	2017 Mar 13	

(continued)

Table 9.1 (continued)

Designation (informal/other name)	Discovered	Period (yrs)	Next perihelion	Notes
173P/Mueller (Mueller 5)	1993 Nov 20	13.6	2021 Dec 16	
174P/Echeclus / 60558 Echeclus	2000 Mar 03	35.3	2015 Apr 22	
175P/Hergenrother (Hergenrother 2)	2000 Feb 04	6.3	2019 Sep 30	
176P/LINEAR / 118401 LINEAR	1999 Sep 07	5.7	2017 Mar 12	
177P/Barnard (Barnard 2)	1889 Jun 24	119.6	2127 Apr 20	
178P/Hug-Bell	1999 Dec 10	7.0	2020 Jul 16	
179P/Jedicke	1995 Jan 09	14.3	2022 May 30	
180P/NEAT	2001 Dec 06	7.5	2015 Dec 12	
181P/Shoemaker-Levy (S-L 6)	1991 Oct 13	7.5	2014 Jun 07	
182P/LONEOS	2001 Oct 26	5.1	2017 Apr 11	
183P/Korlević-Jurić	1999 Feb 18	9.6	2017 Nov 11	
184P/Lovas (Lovas 2)	1986 Jan 09	6.6	2020 Oct 26	
185P/Petriew	2001 Aug 18	5.5	2018 Jan 27	
186P/Garradd	2007 Jan 25	10.6	2019 May 04	
187P/LINEAR	1999 May 12	9.40	2018 May 26	
188P/LINEAR-Mueller	1998 Oct 17	9.1	2017 Feb 17	
189P/NEAT	2002 Aug 03	5.0	2017 Aug 05	
190P/Mueller	1998 Oct 20	8.7	2016 Apr 07	
191P/McNaught	2007 Jul 10	6.6	2014 May 06	
192P/Shoemaker-Levy (S-L 1)	1990 Sep 18	16.4	2024 May 24	
193P/LINEAR-NEAT	2001 Aug 17	6.7	2014 Nov 24	
194P/LINEAR	2000 Jan 27	8.0	2016 Mar 02	
195P/Hill	2006 Nov 22	16.5	2025 Aug 05	
196P/Tichý	2000 Oct 04	7.3	2015 Jun 14	
197P/LINEAR	2003 May 23	4.9	2018 Jan 28	
198P/ODAS	1998 Dec 15	6.8	2018 Dec 14	
199P/Shoemaker (Shoemaker 4)	1994 Oct 14	14.6	2023 Aug 08	
200P/Larsen	1997 Nov 03	10.9	2019 Jul 28	
201P/LONEOS	2001 Sep 10	6.4	2015 Jan 14	
202P/Scotti	2001 Dec 14	7.3	2016 Jun 11	
203P/Korlević	1999 Nov 28	10.0	2020 Mar 05	
204P/LINEAR-NEAT	2001 Oct 13	7.00	2015 Dec 11	
205P/Giacobini	1896 Sep 04	6.7	2015 May 14	
206P/Barnard-Boattini	*1892 Oct 13*	*5.8*	*2014 Aug 27*	
207P/NEAT	2001 Mar 14	7.7	2016 Jul 01	
208P/McMillan	2008 Oct 19	8.1	2016 Jul 01	
209P/LINEAR	2004 Feb 03	5.0	2014 May 06	
210P/Christensen	2003 May 26	5.7	2014 Aug 17	
211P/Hill	2008 Dec 04	6.7	2016 Jan 27	
212P/NEAT	2001 Feb 25	7.8	2016 Sep 10	
213P/Van Ness	2005 Sep 10	6.3	2017 Sep 24	
214P/LINEAR	2002 Feb 07	6.8	2015 Nov 12	
215P/NEAT	2002 May 08	8.1	2019 Nov 18	

(continued)

Table 9.1 (continued)

Designation (informal/other name)	Discovered	Period (yrs)	Next perihelion	Notes
216P/LINEAR	2001 Feb 01	7.7	2016 May 31	
217P/LINEAR	2001 Jun 21	7.8	2017 Jul 16	
218P/LINEAR	2003 Apr 29	6.1	2015 Apr 23	
219P/LINEAR	2002 Jun 05	7.0	2017 Feb 20	
220P/McNaught (McNaught 1)	2004 May 20	5.5	2015 Jun 13	
221P/LINEAR	2002 May 09	6.5	2015 Jul 11	
222P/LINEAR	2004 Dec 07	4.8	2014 Jul 04	
223P/Skiff	2002 Sep 17	8.5	2019 Jan 27	
224P/LINEAR-NEAT	2003 Dec 04	6.3	2016 May 24	
225P/LINEAR	2002 Oct 08	7.0	2016 Aug 16	
226P/Pigott-LINEAR-Kowalski	1783 Nov 19	7.3	2016 Sep 05	
227P/Catalina-LINEAR	2003 Mar 14	6.8	2017 Jun 22	
228P/LINEAR	2001 Dec 17	8.5	2020 Mar 10	
229P/Gibbs	2009 Sep 20	7.8	2017 May 20	
230P/LINEAR	2009 Oct 27	6.3	2015 Nov 18	
231P/LINEAR-NEAT	2003 Apr 29	8.1	2019 Jun14	
232P/Hill	1999 Nov 18	9.5	2019 Apr 06	
233P/La Sagra	2009 Nov 19	5.3	2015 Jun 25	
234P/LINEAR	2002 Feb 08	7.5	2017 Jun 01	
235P/LINEAR	2002 Mar 16	8.0	2018 Mar 17	
236P/LINEAR	2003 Oct 29	7.2	2017 Nov 20	
237P/LINEAR	2003 Mar 14	7.2	2016 Oct 11	
238P/Read	2005 Jul 27	5.6	2016 Oct 22	
239P/LINEAR	1999 Dec 07	9.5	2019 Jan 10	
240P/NEAT	2003 Mar 23	7.6	2018 May 16	
241P/LINEAR	1999 Oct 30	11.0	2021 Jul 25	
242P/Spahr	1999 Oct 27	13.0	2024 Dec 23	
243P/NEAT	2003 Sep 24	7.5	2018 Aug 26	
244P/Scotti	2000 Dec 30	10.8	2022 Nov 18	
245P/WISE	2010 Jun 02	8.0	2018 Feb 08	
246P/NEAT	2004 Mar 28	8.1	2021 Feb 22	
247P/LINEAR	2002 Nov 05	7.9	2018 Dec 02	
248P/Gibbs	2010 Nov 27	14.6	2025 Sep 14	
249P/LINEAR	2006 Oct 19	4.6	2015 Nov 26	
250P/Larson	2011 Jan 10	7.2	2018 Feb 02	
251P/LINEAR	2004 Apr 17	6.5	2017 Jul 16	
252P/LINEAR	2000 Apr 07	5.3	2016 Mar 15	
253P/PanSTARRS	2011 Sep 04	6.5	2018 May 07	
254P/McNaught	2010 Oct 04	10.1	2020 Sep 29	
255P/Levy	2006 Oct 02	5.3	2017 May 03	
256P/LINEAR	2003 Apr 26	10.0	2023 Mar 12	
257P/Catalina	2003 Apr 26	7.3	2020 Sep 10	
258P/PanSTARRS	2012 Apr 27	9.2	2020 Jun 19	

(continued)

Table 9.1 (continued)

Designation (informal/other name)	Discovered	Period (yrs)	Next perihelion	Notes
259P/Garradd	2008 Sep 02	4.5	2017 Aug 04	
260P/McNaught	2005 May 20	7.1	2019 Sep 09	
261P/Larson	2005 Jun 07	6.8	2019 Jun 18	
262P/McNaught-Russell	1994 Dec 12	18.3	2030 Dec 15	
263P/Gibbs	2006 Dec 28	5.4	2017 Sep 29	
264P/Larsen	2004 Apr 22	7.7	2019 Aug 04	
265P/LINEAR	2003 Jul 30	8.7	2021 Feb 09	
266P/Christensen	2006 Oct 27	6.6	2020 Apr 19	
267P/LONEOS	2006 Aug 29	6.0	2018 Jul 22	
268P/Bernardi	2005 Nov 01	9.8	2015 Apr 27	
269P/Jedicke	1996 Jan 14	19.8	2014 Nov 14	
270P/Gehrels	1997 Feb 01	18.0	2030 Aug 17	
271P/van Houten-Lemmon	1960 Sep 24	18.4	2032 Mar 22	
272P/NEAT	2004 Mar 18	9.4	2022 Jul 17	
273P/Pons-Gambart	1827 Jun 21	188.1	2191 Jun 16	
274P/Tombaugh-Tenagra	2012 Nov 27	9.1	2022 Apr 081	
275P/Hermann	1999 Feb 18	13.8	2026 Oct 27	
276P/Vorobjov	2012 Oct 15	12.5	2024 Dec 10	
277P/LINEAR	2005 Nov 03	7.6	2020 Dec 30	
278P/McNaught	2006 Jun 27	7.1	2020 Sep 12	
279P/La Sagra	2009 Oct 10	6.8	2016 Jul 14	
280P/Larsen	2004 May 10	9.6	2013 Dec 10	
281P/MOSS	2000 Nov 17	10.7	2023 Feb 01	
282P/(323137)	2003 Jan 31	8.8	2021 Oct 24	
283P/Spacewatch	2013 Apr 07	8.4	2021 Sep 08	
284P/McNaught	2007 Aug 17	7.0	2014 Sep 02	
285P/LINEAR	2003 Dec 04	9.5	2023 Jan 12	
286P/Christensen	2005 Aug 24	8.4	2014 Jan 06	
287P/Christensen	2006 Jun 15	8.5	2014 Dec 17	
288P/(300163)	2006 Mar 21	5.3	2016 Nov 10	
289P/Blanpain	1819 Nov 20	5.3	2019 Dec 20	
290P/Jäger	**1998 Oct 24**	**15.2**	**2014 Mar 12**	
291P/NEAT	2004 Mar 27	9.7	2014 Dec 18	

Bold indicates comets described in more detail in this chapter
Italics for comets mentioned elsewhere in book

Appendix A

Comets Through the Eyes of Space Probes

Our ideas about small Solar System bodies (SSSBs)—asteroids, planetary moons and cometary nuclei—have been dramatically transformed over the past few decades by a variety of space probes that have flown by, imaged, landed on and even sampled their material. In addition, tremendous advances in remote sensing and imaging from Earth-based and Earth-orbiting telescopes have enabled these objects to be studied in great detail. CCD imaging has led to comets to be pictured in unprecedented detail, observations in other wavelengths has told us a great deal about their temperature and composition, and radar has been able to create high-resolution pictures of the surfaces of objects that have approached near to Earth.

As we have seen, SSSBs are part of a broad continuum of Solar System bodies. Some main belt asteroids display comet-like activity. Numerous small satellites around Jupiter, Saturn, Uranus and Neptune are undoubtedly captured comets, and the icy worlds of the Kuiper Belt are simply large cometary nuclei, inactive due to their great distance from the Sun. Gaining a better understanding of these objects through robotic reconnaissance helps astronomers build up a better picture of SSSBs and the Solar System in general.

Jupiter Takes Its Knocks

In the summer of 1994, one of the most spectacular astronomical events to occur in our Solar System for many hundreds of years astonished astronomers the world over. From July 16–22, more than 21 pieces of the fragmented nucleus of comet P/

P. Grego, *Blazing a Ghostly Trail: ISON and Great Comets of the Past and Future*,
The Patrick Moore Practical Astronomy Series, DOI 10.1007/978-3-319-01775-4,
© Springer International Publishing Switzerland 2014

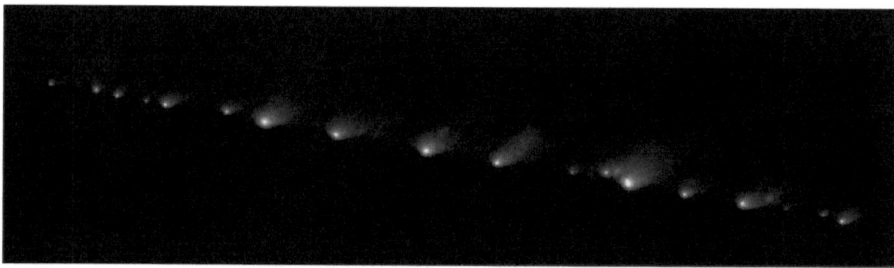

Fig. A.1 The fragmented comet P/Shoemaker-Levy 9, imaged in May 1994 by the Hubble Space Telescope, 2 months before its collision with Jupiter. (NASA/STScI)

Shoemaker-Levy 9—ranging in size from lumps 600 m across to the largest, measuring more than 4 km in diameter—slammed one by one into the atmosphere of the giant planet Jupiter.

Comet P/Shoemaker-Levy 9 (SL9) was discovered in March 1993 by American astronomers Eugene and Carolyn Shoemaker and David Levy; the object represented the team's ninth joint comet discovery. After careful observations, it was calculated that the comet had once been in a distant near-circular orbit around the Sun in the ecliptic plane between the orbits of Jupiter and Saturn. Around the year 1929 (the very year of Carolyn Shoemaker's birth), SL9 was captured by Jupiter's gravity, and it assumed a near-polar orbit around the giant planet, making one revolution around Jupiter in 2–2.5 years. The comet made several close approaches to Jupiter in the ensuing years, notably in 1940–1942 and in 1970. In July 1992, 8 months before its discovery, the comet approached the cloud tops of Jupiter to a distance of only 43,000 km, ten times closer than the innermost Galilean satellite Io. Consequently, the immense gravitational forces SL9 endured during this very close pass were enough to break the single icy nucleus into many smaller pieces.

When SL9 was closely imaged it assumed a peculiar string-of-pearls appearance in the midst of the faint haze of its dusty coma, a mass of non-resolvable debris arising from the fragmentation. No gaseous emissions were detected from the comet, which did not surprise scientists, since the solar energy received by SL9 at that distance was too weak to produce noticeable activity (Fig. A.1).

Many months before the fragments of SL9's nucleus were predicted to hit Jupiter's southern hemisphere, astronomers began to speculate about how easy it would be to see the effects of the impacts. The entry of each cometary nucleus into the Jovian atmosphere would be preceded by a magnificent coma 'meteor' storm, as billions of dust particles burned up high above Jupiter's clouds. From Earth each storm would be observed as a flaring event on Jupiter's limb that would last a few seconds. The impacts themselves would not be directly visible from Earth, since the cometary fragments would arrive on Jupiter's far-side, just past the visible edge of the planet. However, the rapid rotation of Jupiter (the giant planet's day is less than

10 h long) would quickly bring the impact sites into view. If there were any scars visible in Jupiter's clouds then they would rotate on to the Jovian near side in less than half an hour after their formation.

Those who saw the effects of SL9's smash into the clouds of Jupiter witnessed a spectacle the likes of which we are unlikely to see again for a very long time. It was an awe-inspiring sight that amateur astronomers will never forget. On the evening of July 20 the most incredible sight presented itself in the field of view of even small telescopes. Mighty Jupiter—the biggest planet in the Solar System—appeared visibly scarred by fragments of impacting cometary nuclei. These scars took the form of intense black spots surrounded by shock wave patterns. The most prominent scar, caused by the impact of 4-km-diameter Fragment G, took the form of a black blob bordered on its preceding edge and to the south by a semicircle of dark material. Over the next few days the scene on Jupiter constantly changed before the eyes of astronomers, as further impacts created new atmospheric disturbances and old ones developed into complicated shapes in the southern temperate region.

Astronomers eagerly followed Jupiter into the evening twilight over the months following the SL9 impacts as the planet headed for conjunction with the Sun in November. When Jupiter had rounded the Sun and emerged into December's morning skies, the effects of the impacts that happened 6 months before were still noticeable through small telescopes. Material that had been brought to the cloud tops from deep within the Jovian atmosphere had spread around the latitude of the impacts, and some traces of the old impact sites were still detectable within this new dusky 'south-south temperate belt' (Fig. A.2)

Research into old astronomical archives brought to light several examples of unusual Jovian atmospheric phenomena observed throughout history. In December 1690 Giovanni Cassini observed the sudden appearance of a single black spot in Jupiter's equatorial region that, over the next 18 days, developed in complexity and spread over a portion of that latitude.

On July 19, 2009, Anthony Wesley, an Australian astrophotographer, imaged a new dark marking that suddenly appeared on Jupiter. The marking, in the planet's south polar region, appeared remarkably similar to those produced when Comet Shoemaker-Levy 9 crashed into the giant planet in 1994. Many other imagers and observers, the author included, confirmed the feature (Fig. A.3).

Without warning on June 3, 2010, a small asteroid or comet collided with Jupiter, producing a brief flash of light that was captured by the CCD cameras of Anthony Wesley (imager of the 2009 impact spot on Jupiter) and Christopher Go in the Philippines. The impact took place near the edge of the giant planet at the southern border of its much-faded south equatorial belt. Despite a speedily issued alert, follow-up images and observations failed to confirm any obvious disruption to Jupiter's atmosphere, such as the dark spots resulting from the larger impacts of 1992 and 2009 (Fig. A.4).

Just 11 weeks later, on August 20, 2010, Jupiter was once again impacted by a small asteroid or comet, indicating that the rate of observable Jovian impacts is far greater than previously thought. The impact flash was captured by Masayuki

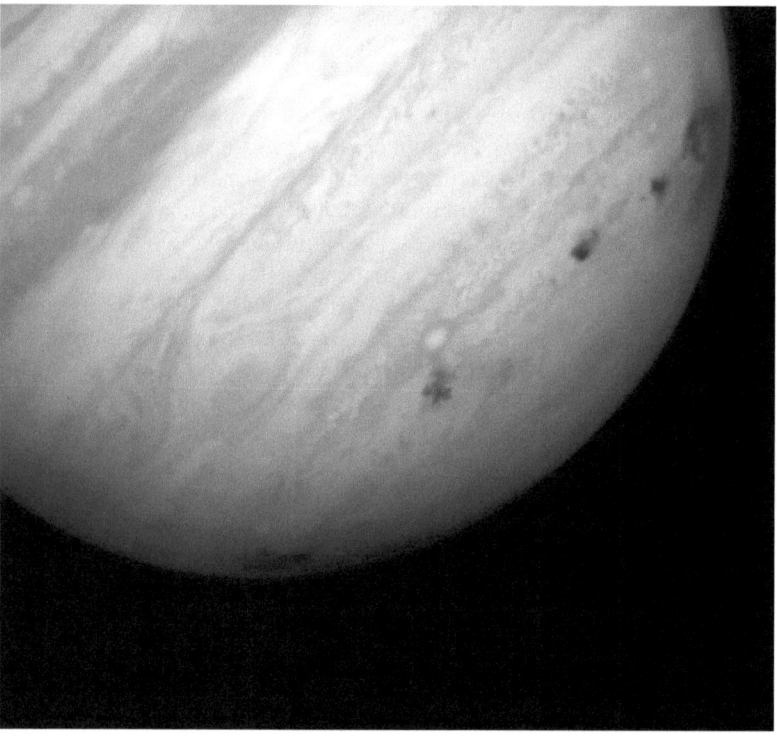

Fig. A.2 Scars in Jupiter's atmosphere caused by the impact of the fragmented nucleus of comet SL9 in July 1994. (NASA/STScI)

Tachikawa from Kumamato, Japan, who used a ToUcam Pro II (a webcam) attached to a 6-in. apochromat and 5× Barlow lens to secure a routine video of Jupiter. Later examination of the video revealed a brief flash of light at 18:22 UT, as the fireball flared in Jupiter's atmosphere above the north edge of the north equatorial belt. The flash was independently confirmed, but like the previous one, it left no trace of a disturbance in Jupiter's clouds (Fig. A.5).

Close-Up of a Disintegrating Comet

C/1999S4 (LINEAR) was discovered on September 27, 1999, by the Lincoln Laboratory Near Earth Asteroid Research (LINEAR) program, a U.S. study designed to spot all asteroids and comets that come dangerously close to Earth. The comet, fresh from the Oort Cloud and making its first trip into the inner Solar System, passed through northern skies and near the celestial pole in mid-March 2000 and became a naked-eye object in July. Professional astronomers noticed that the comet's brightness was rather variable, and it also showed large 'non-gravitational'

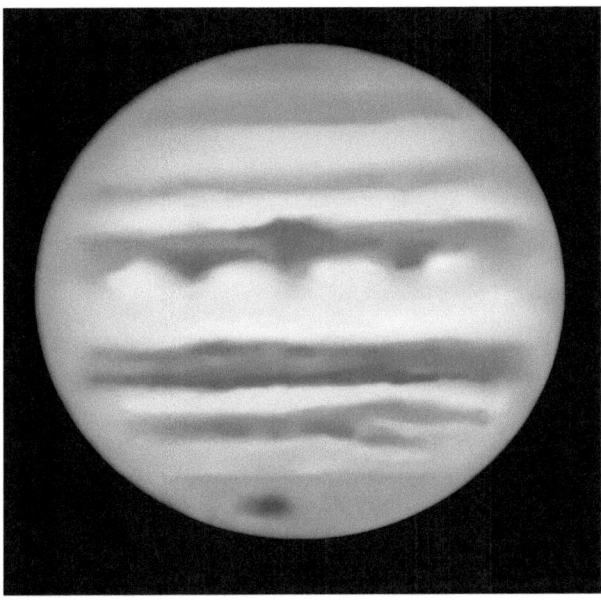

Fig. A.3 Author's observation of the impact scar left in Jupiter's south polar region, made on July 29, 2009, at 00:15 UT using an 8-in. SCT (180×)

effects that caused it to drift slightly from its predicted path. Jetting of material from an active nucleus can produce thrust that creates such an effect.

An outburst and a minor fragmentation event took place on July 5, followed by a stronger outburst on July 20–21, 5 days before perihelion; the comet then faded dramatically, its coma shrinking and becoming more diffuse, while its tail faded away. Images taken by the Hubble Space Telescope and the ESO VLT showed the cause. The 900-m diameter nucleus of C/1999S4 had completely fragmented, disintegrating into a myriad of tiny cometesimals, the largest of which were around 100 m across. It was the first comet ever to have been caught in the very act of complete disintegration. By early August the comet was a dim smudge that rapidly faded from view (Fig. A.6).

Radar Captures Kleopatra

Asteroid 216 Kleopatra was discovered photographically by Johann Palisa from Pula, Croatia, on April 10, 1880. At opposition it shines at about 12th magnitude and can be glimpsed as a star-like point through amateur instruments. In September 2008 images taken by the Keck Observatory in Mauna Kea, Hawaii, showed that Kleopatra had two small moons of its own; they were named Alexhelios and Cleoselene, after Cleopatra's twin children.

Jupiter + Fireball

Anthony Wesley, Broken Hill Australia
3 Jun 2010 20:31.6 Z   CMI 299   CMII 33   CMIII 209

Fig. A.4 The Jupiter impact flash of June 3, 2010, imaged by Anthony Wesley. (Used with permission)

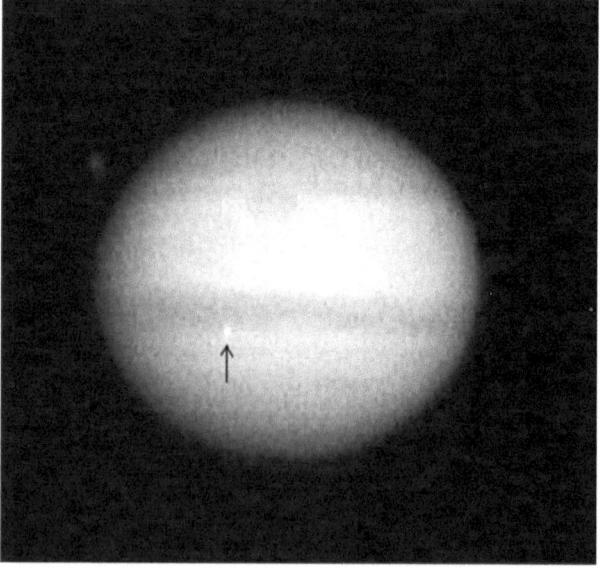

Fig. A.5 The Jupiter impact flash of August 20, 2010, imaged by Masayuki Tachikawa. (Used with permission)

Fig. A.6 The dramatic brightening and fragmentation of C/1999S4 (LINEAR) on August 5, 2000, imaged from the ground and in more detail by the Hubble Space Telescope. (NASA/H. Weaver, Johns Hopkins University.)

In November 1999 a team of American astronomers used the famous 305-m radio telescope at Arecibo in Puerto Rico to bounce radar signals off Kleopatra when the asteroid was 171 million km from Earth, and using sophisticated computer analysis the echoes were decoded and transformed into a computer model of the asteroid's shape. These first detailed images of a main belt asteroid revealed it to be a 217-km-long fragment of space debris shaped remarkably like a giant dog bone. The unusual shape may have resulted from a collision between two asteroids that did not completely shatter and disperse all the fragments. The radar observations indicate that Kleopatra is highly metallic, with a porous and loosely consolidated surface a bit like the surface of the Moon, although of a different composition (Fig. A.7).

Splitting of C/2001 A2 (LINEAR)

C/2001 A2 was discovered by the LINEAR. At the end of March 2001, as it headed towards a good showing in northern skies, C/2001 A2 (LINEAR) was observed to break into two. When the nucleus split, fresh volatile ices inside were

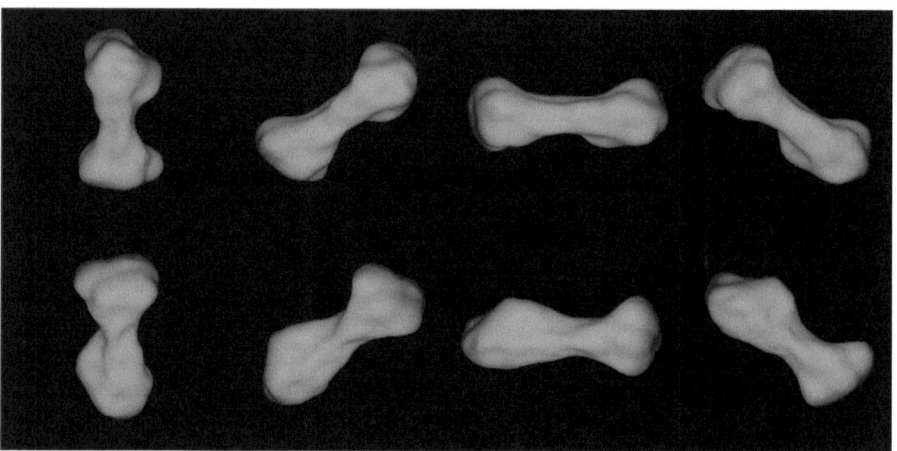

Fig. A.7 These radar images, taken by the Arecibo radio telescope, show the rotation of bone-shaped asteroid Kleopatra. (NASA)

suddenly exposed to solar heating and sublimated to gas, which in turn liberated dust particles trapped in the ices. Increasing dramatically in brightness, the comet flared from magnitude 13 to 8—a hundred fold increase. By April 25 the comet could be seen with a keen naked eye in dark skies, and on April 30 the 1.5-m Catalina Telescope on Mount Bigelow, Arizona, successfully imaged the split nucleus (Figs. A.8 and A.9).

Comets Blasted by the Sun

Between February 17 and 20, 2003, comet C/2002V1 (NEAT) became the brightest comet ever observed by the Large Angle and Spectrometric Coronagraph (LASCO) aboard the solar satellite SOHO (Solar and Heliospheric Observatory) operated by NASA and ESA. As the comet swept past the Sun at a distance of 14.8 million km, its bright, broad tail was scrambled by two coronal mass ejections (CMEs) in the sort of interaction that had never been seen before. The encounter split the ion tail, one part of which stayed with the nucleus, while the other was captured by the magnetic field of the CME and swept into the dust tail (Fig. A.10).

Nearly 4 years later, in mid-January 2007, comet C/2006 P1 (McNaught) came within LASCO's field of view—not only did it become even brighter than C/2002V1 (NEAT), it was visually the brightest comet observed in over 40 years. As C/2006 P1 swung past the Sun between January 12 and 15—a perihelion distance of just 0.17 au—it brightened tremendously, even without the aid of any CMEs (Fig. A.11).

Fig. A.8 The split nucleus of C/2001 A2 (LINEAR), imaged on April 30, 2001, with the 1.5-m Catalina telescope, Mount Bigelow. (C. W. Hergenrother, M. Chamberlin and Y. Chamberlain.)

Juno Revealed

Demonstrating just how far CCD imaging had progressed, in August 2003 high resolution images of the surface of asteroid 3, Juno, taken with the famous 100-in. Hooker telescope at Mt. Wilson Observatory, United States, revealed that it had a large crater 'bitten' out of its side. About 100 km across, the crater is about half the diameter of the asteroid itself. Sallie Baliunas and colleagues imaged Juno using an adaptive optics system when it was just 1.25 au (190 million km) from Earth. Even this close, Juno appeared very tiny—just half an arcsecond across (Fig. A.12).

73P Is Breaking Up

Comet 73P/Schwassmann-Wachmann was discovered on photographic plates taken at the Hamburg Observatory, Germany, on May 7, 1930. That year it approached Earth at a distance of 0.06 au (9.3 million km), making it the 15th closest cometary approach to our planet. It was found to be a periodic comet with a 5.36-year orbit

Fig. A.9 C/2001 A2 (LINEAR) was observed and drawn by the author on July 14, 2001, using a 4-in. Maksutov (30×)

(coming near Earth every 16 years), and being the 73rd known periodic comet, it was allocated the designation 73P. It was also Arnold Schwassmann's and Arno Wachmann's third joint discovery, so the comet is sometimes known as Schwassmann-Wachmann 3.

On the return of 73P to perihelion in 1995, astronomers weren't expecting to see much more than a 12th magnitude object; instead, the comet was found to be far brighter than expected, in September and October reaching naked-eye visibility and sporting a tail. The comet was observed to undergo two outbursts as its nucleus broke into four hefty components, labeled 73P-A to D, of which C was the largest.

During its close approach to Earth of 0.07 au (10.5 million km) in the spring of 2006 the comet once again proved to be visible with the unaided eye. Two of the fragments—73P-B and the main component, 73P-C—were easily visible through amateur telescopes, the latter even growing a short tail. Larger instruments revealed other, fainter fragments with erratic light curves. In March 2006 at least eight fragments were known, namely 73P-B, C, G, H, J, L, M and N. On April 18, 2006, the Hubble Telescope revealed dozens of fragments, and it is known that the comet had split into at least 66 pieces.

73P has already been linked with the minor Tau Herculid meteor shower (active between May 19 to June 19 each year). Since 73P's orbit approaches so close to Earth, the recent fragmentation has enriched the meteoroid stream considerably,

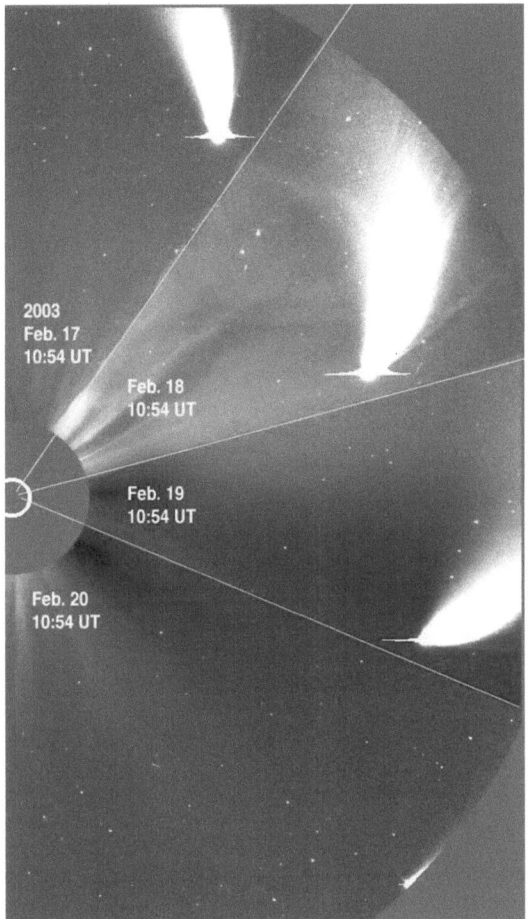

Fig. A.10 C/2002V1 (NEAT), imaged in February 2003 by LASCO on the SOHO satellite. (NASA/ESA)

so any associated meteor activity in the future will probably take the form of a short-lived outburst in years when the comet is at perihelion and close to Earth, but his won't take place until 2022 (Figs. A.13 and A.14).

Comet Lulin Goes Green

Making its first visit to the inner Solar System, comet C/2007N3 (Lulin) was discovered photographically in 2007 by astronomers Ye Quanzhi and Lin Chi-Sheng from Taiwan's Lulin Observatory. It approached to within 0.4 au (61 million km)

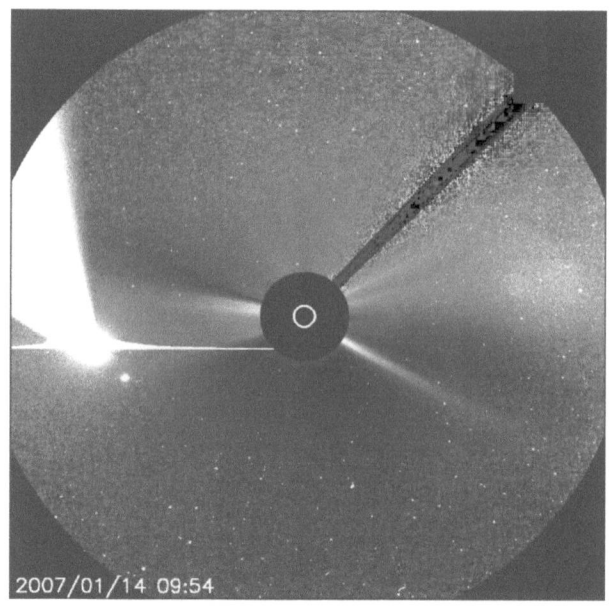

Fig. A.11 Comet McNaught, imaged by LASCO on the SOHO satellite on January 14, 2007. (NASA/ESA)

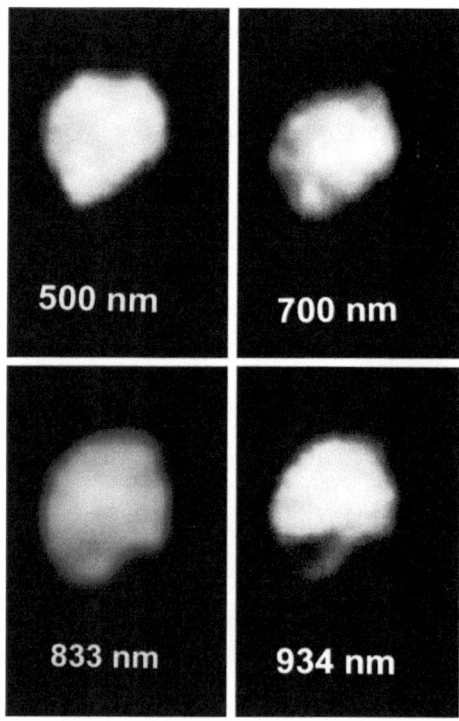

Fig. A.12 Main belt asteroid Juno, imaged in different wavelengths with the Hooker Telescope at Mt. Wilson. (© Sallie Baliunas et al.)

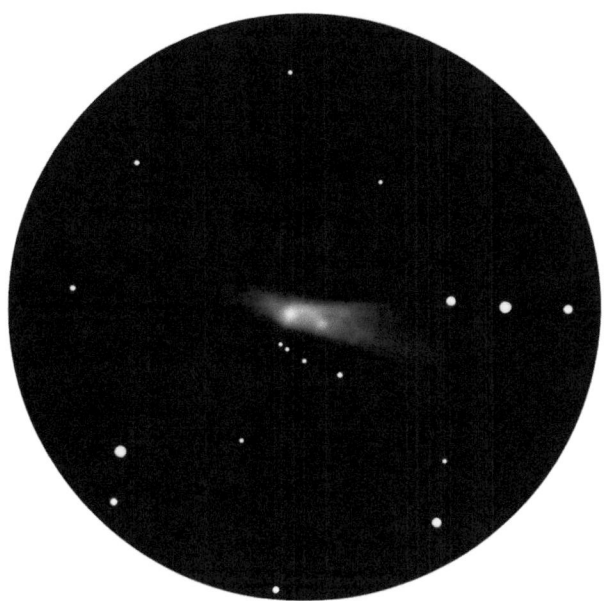

Fig. A.13 Using 15×70 binoculars, the author made this observational drawing of fragment B on May 6, 2006

of Earth on February 23, 2009, becoming visible with the unaided eye as it tracked through Leo and Cancer. Observers were treated to the splendid sight of the comet passing 2° from Saturn on the evening of perigee, plus its close approach to the large open star cluster M44 on March 5/6. What gave C/2007N3 even more of a visual appeal was its development of a short tail and anti-tail, both of which were visible through binoculars, and the distinctly green hue of its coma, produced by the presence of carbon and cyanogen.

NASA's Swift satellite took the opportunity to view C/2007N3 using its Ultraviolet/Optical and X-Ray telescopes. Observations revealed that the nucleus was shedding around 10,000 cubic meters of water an hour. That's more than 3 % of the average flow rate of the River Thames through London (Figs. A.15 And A.16).

Hubble Images Asteroid Smash

Astronomers initially thought that P/2010 A2 (LINEAR), discovered on January 6, 2010, was a main belt comet, hence its cometary designation. It had a bright nucleus and sported a tail, and had gone through perihelion the previous month.

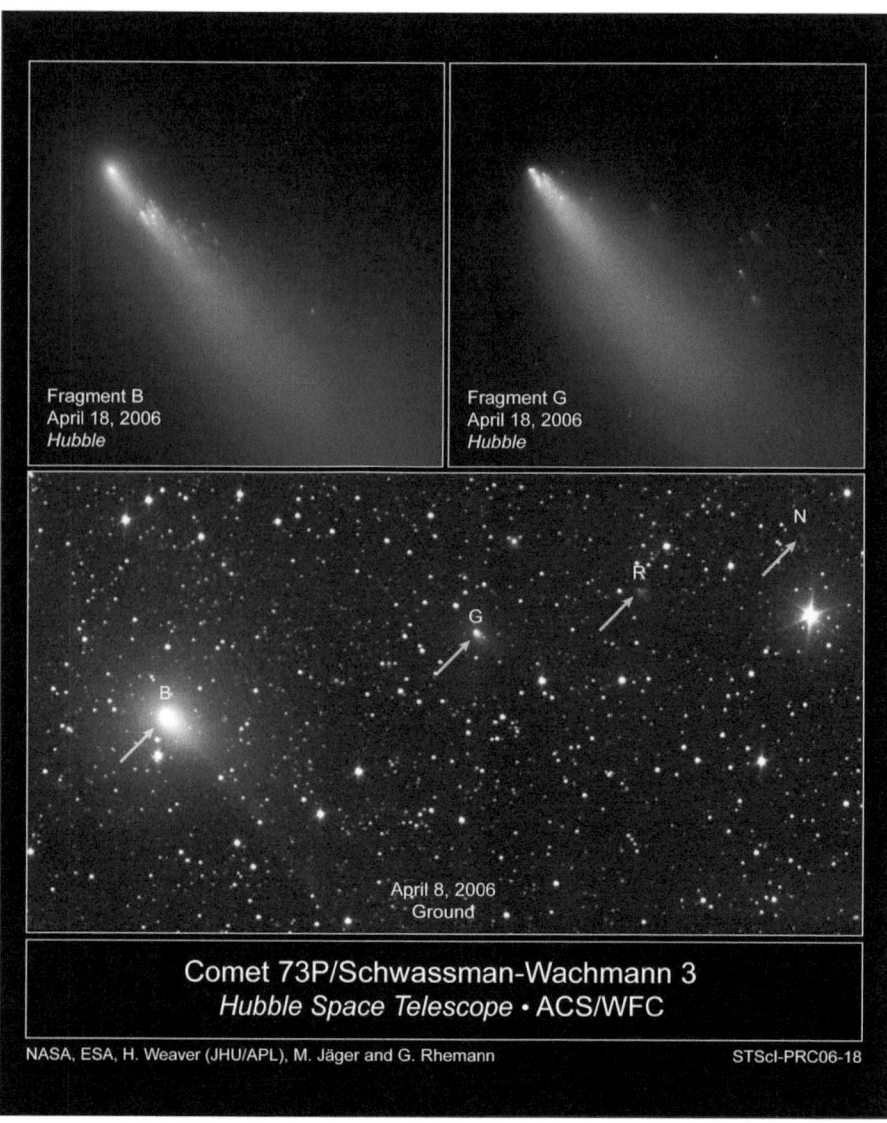

Fragment B
April 18, 2006
Hubble

Fragment G
April 18, 2006
Hubble

N

R

G

B

April 8, 2006
Ground

Comet 73P/Schwassman-Wachmann 3
Hubble Space Telescope • ACS/WFC

NASA, ESA, H. Weaver (JHU/APL), M. Jäger and G. Rhemann   STScI-PRC06-18

Fig. A.14 The break-up of 73P/Schwassman-Wachmann 3, imaged by the Hubble Space Telescope in April 2006. (NASA)

Within a month after its discovery, after the Hubble Space Telescope had imaged it in great detail, it became clear that this was no comet. Hubble's images showed an object surrounded by debris and trailing streamers of dust. The main nucleus of P/2010 A2—estimated to be some 140 m across—was clearly shown to lie outside

Fig. A.15 Comet Lulin, observed by the author at 01:45 UT on February 23, 2009, using 25 × 100 binoculars

Fig. A.16 Comet Lulin, imaged by NASA's Swift satellite as it made it closest approach to Earth in February 2009. (NASA/STScI)

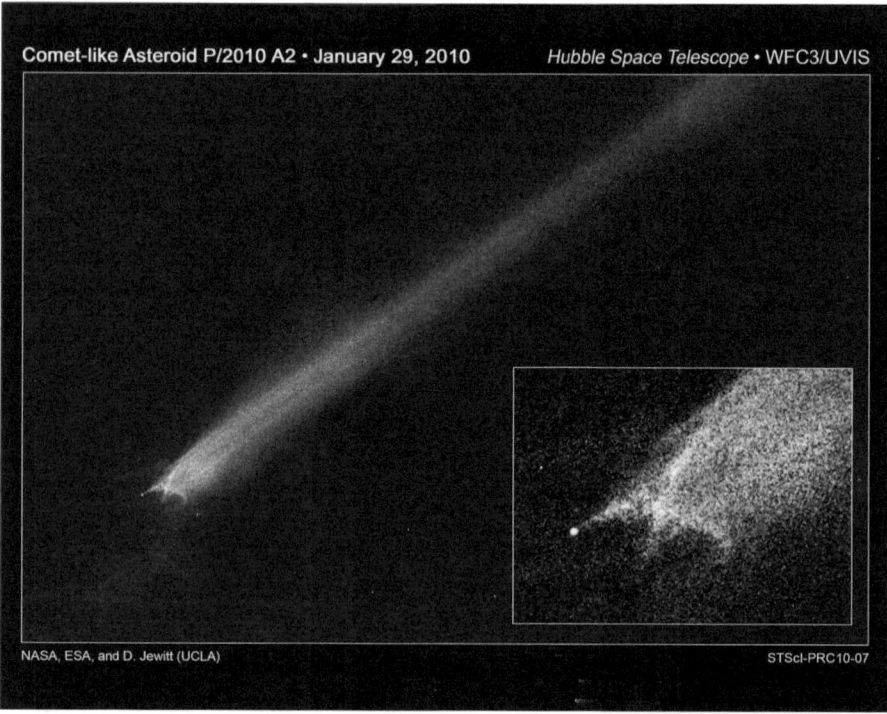

Comet-like Asteroid P/2010 A2 · January 29, 2010   *Hubble Space Telescope* · WFC3/UVIS

NASA, ESA, and D. Jewitt (UCLA)   STScI-PRC10-07

Fig. A.17 The comet-like appearance of P/2010 A2—debris thrown off by a collision in deep space—imaged by the Hubble Space Telescope on January 29, 2010. (NASA)

these filaments of dust. It was quite unlike a comet's tail, and indications were that a collision between two asteroids has taken place. It was the first time that an asteroid collision had been imaged. It's likely that the complex trail of debris resulted from a head-on impact a year earlier between two small and previously unknown asteroids traveling at speeds 5 times faster than a rifle bullet. Like the tail of a comet, the pressure of sunlight was sweeping the debris into a tail, away from the surviving remnant of the collision. The orbit of P/2010 A2 places it within the Flora asteroid family, itself produced by a collisional shattering more than 100 million years ago (Fig. A.17).

Space Probe Views of Interplanetary Rock and Ice

Incredible images of asteroids and comets have been secured by space probes dispatched by NASA, ESA, JAXA and CNSA (the Chinese Space Agency), providing plenty of information with which to refine our understanding of the history of the Solar System and the processes that have shaped its objects.

Halley's Comet

VeGa 1, VeGa 2, Sakigake, Suisei and Giotto—a small armada of unmanned spacecraft constructed by Russia, Europe and Japan—was sent to intercept Halley's Comet in 1986. In March 1986 the two Soviet VeGa probes and the European Space Agency's Giotto craft passed through Halley's coma. On March 4 and 6, respectively, VeGas 1 and 2 passed within 8,890 and 8,030 km of Halley's nucleus, taking scores of low-resolution images as they sped by.

On March 14 the Giotto probe approached the nucleus to a distance of just 605 km and took many stunning high-resolution images. One problem the Giotto engineers had to attempt to minimize was the damage caused by impacting dust grains as the probe approached the comet at an incredible relative speed of 68 km per second—50 times the speed of a rifle bullet. It was decided to give the probe a double dust bumper on its leading end and spin it around its long axis once every 4 s in order to stabilize it in flight. Special 'nutation dampers' could be activated to cancel out any undue wobble in the craft's motion that might be caused by a small impact.

Giotto's multicolor camera was mounted at its forward end, peeking at a shatterproof metal mirror through a cutaway in the dust bumpers. The spin of the craft meant that the images were made up of circular sweeps that took in a section of the coma and was aligned by an onboard computer to include the brightest part of the comet's nucleus on each sweep. An ingenious experiment was devised to detect the size and locations of impacts of small dust grains on the dust bumper. Some scientists gave the probe a 1-in-10 chance of being destroyed completely as it headed towards Halley's nucleus—not exactly a kamikaze venture, more a fighting chance of survival in a single-shot game of Russian roulette.

Because Giotto's camera kept the brightest part of the scene in its field of view, images of the dark nucleus that gave rise to these bright jets of dust and gas were visible only when the craft was at some distance from the nucleus. The last images containing views of the dark nucleus were taken about 50 s prior to closest approach, a distance of some 3,500 km. Further in, the images show just the bright nuclear jets and a certain amount of fine structure within them.

Millions watched the final phase of the Giotto approach near Halley's heart, as the event was covered live on television. Viewers watched excitedly as monitors flashed a rapid succession of false-color images transmitted by the large Earth-pointing parabolic dish attached to the rear of the distant spacecraft. In all, the probe returned more than 2,000 pictures of Halley.

The first dust particle to impact on Giotto's bumper was felt when the probe was some 287,000 km from the nucleus. Between a distance of 150,000 and 90,000 km there was a steady pitter-patter of small impacts, after which the rates fell until the probe was 70,000 km distant from the nucleus, when activity then began to steadily increase. The amount of dust detected in these outlying regions of the coma had been predicted to be far higher.

However, just as astronomers began to think that Halley's coma was a relatively dust-scarce zone, 20 s prior to the probe's closest approach the dust impact rate went off the measurable scale, with the thickest zone of dust encountered in the last minute before closest approach. More than a hundred of the larger-sized dust particles in the last stages of approach managed to bump Giotto into a temporarily unrecoverable attitude. The last images secured by Giotto were taken at a distance of 1,350 km from Halley's nucleus and showed a fan-shaped jet region about 15 km across.

Halley's nucleus was found to be a dark, elongated, potato-shaped body around 15 km long and 8 km wide. Later image enhancements have revealed a degree of topographic form, including scalloped ridges, hills and several craters. Some of the craters may be impact scars, while others were undoubtedly areas excavated when pockets of volatiles under Halley's crust exploded.

With 60 kg of fuel remaining after the Halley encounter, it was decided to send Giotto on to study comet 26P/Grigg-Skjellerup. In July 1992, Giotto crossed the comet's bow shock and entered the coma about 17,000 km from the nucleus, towards which it had been directly aimed, ending up missing the nucleus by just 200 km. Unfortunately no images were returned because its camera had been put out of commission during the encounter with Halley, but through Giotto's remaining operating instruments much was learned about the interactions between the comet, the solar wind and the interplanetary magnetic field (Fig. A.18).

Galileo Glides By

951 Gaspra was the first asteroid to come under close scrutiny when, on October 29, 1991, the Galileo probe (en route to Jupiter) flew by it at a distance of 1,600 km at a relative speed of 29,000 km per hour. The best images have a resolution of around 50 m per pixel. Gaspra was found to be a rocky object measuring $18 \times 11 \times 9$ km (about the same size as the Martian satellite Phobos) and covered with small impact craters, some showing ejecta systems. Grooves some 100–300 m wide and up to 2½ km long run across the asteroid's surface; these may have resulted from a large impact in the past and indicate that Gaspra is a solid object, a fragment of a shattered parent asteroid, rather than a loose rubble pile (Fig. A.19).

Following a mid-course trajectory change, Galileo went on to encounter asteroid 243 Ida on August 28, 1993. Flying by at a distance of 2,390 km and a relative velocity of 49,000 km per hour, Ida was found to be elongated, with an average diameter of 31 km. Heavily cratered, it shows a long history of meteoroidal bombardment, with many generations of craters visible. Like Gaspra, Ida displays grooves indicating that it was once part of a larger body that experienced a catastrophic fragmentation. Analysis of the images soon after the flyby showed that Ida has a tiny moon, Dactyl, just 1.4 km across (Fig. A.20).

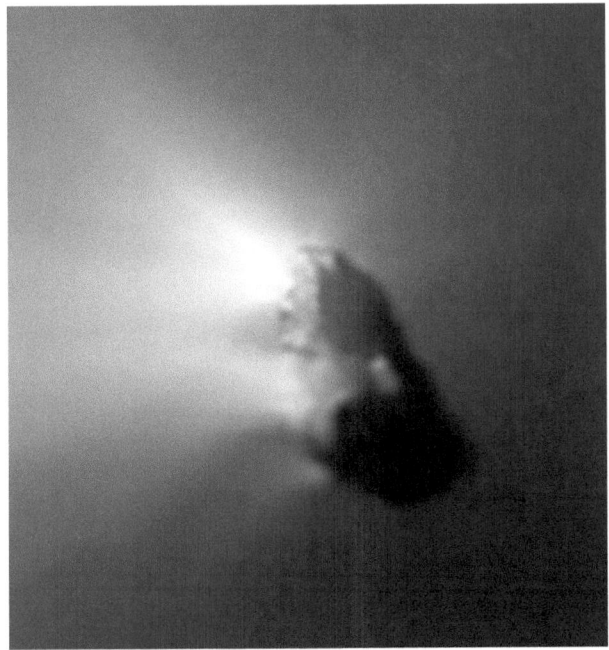

Fig. A.18 The nucleus of Halley's Comet, imaged in March 1986 by ESA's Giotto probe. (ESA)

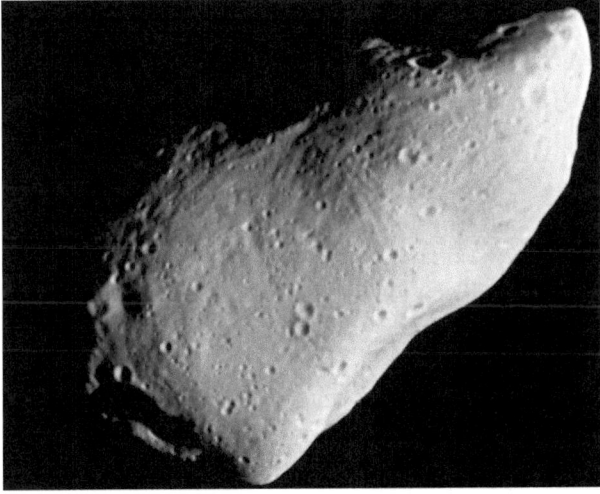

Fig. A.19 Asteroid Gaspra, imaged in October 1991 by the Galileo probe. (NASA/JPL)

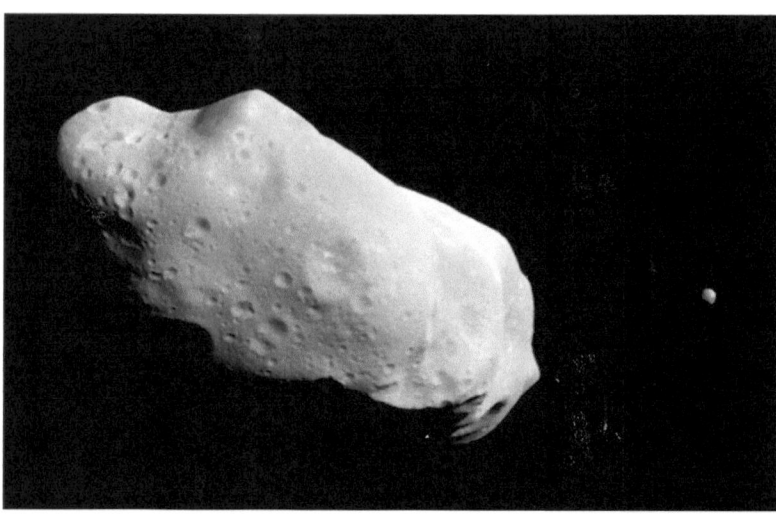

Fig. A.20 Asteroid Ida and its tiny moon Dactyl, imaged in August 1993 by the Galileo probe. (NASA/JPL)

Views from NEAR-Shoemaker

NASA's next big rendezvous took place on June 27, 1997, when the NEAR-Shoemaker space probe, en route to asteroid Eros, flew within 1,200 km of asteroid 253 Mathilde. Measuring 66×48×46 km, Mathilde is a very dark minor planet with an intensely cratered surface. The pitch-dark asteroid is dented with five craters bigger than 20 km across, the largest of which, named Karoo, is more than half the asteroid's diameter and around 10 km deep—proportionally, the largest impact feature on any object in the Solar System. Just how Mathilde survived intact after experiencing such a battering remains a mystery—it should have been smashed to bits—but the fact that it has a relatively low density indicates that it is a loosely packed rubble pile that would have distributed the shock of impact less efficiently (Fig. A.21).

On February 14, 2000, the NEAR-Shoemaker probe finally reached its destination, asteroid 433 Eros, becoming the first probe to orbit an asteroid. Eros, a Mars-crossing asteroid, was found to measure 34×11×11 km; like other asteroids, it is heavily cratered. During the time that the craft was in orbit, its height above the surface was raised and lowered in increments; from its original 200-km orbit it was brought down to 50 km (April) and then 35 km (July), raised again to 200 km and finally back to 35 km by December 2000. In January 2001 the craft made a series of close passes to Eros, coming within 3 km of its surface.

Eros was found to be strewn with large boulders, most of which were determined to have been ejected from a single crater (designated Charlois Regio) formed by an impact around a billion years ago. Shock waves from the impact turned older

Fig. A.21 Asteroid Mathilde, imaged in June 1997 by the NEAR-Shoemaker probe. (NASA/JPL)

Fig. A.22 Asteroid Eros, imaged in 2000 by the orbiting NEAR-Shoemaker probe. (NASA/JPL)

craters around it into rubble, and a ridge called Hinks Dorsum is a thrust fault caused by the impact (Fig. A.22).

On February 12, 2001, the NEAR-Shoemaker spacecraft made a successful landing on Eros. What made this feat even more amazing was that the probe was not originally built to land, merely to study it from orbit. Having fulfilled all its

NEAR's landing site on asteroid
Eros (arrowed). Two images from
NEAR's descent, taken from 250
netres and (right) 120 metres, where
transmission was interrupted as the
probe landed. Credit: NASA/JPL.

Fig. A.23 NEAR-Shoemaker's landing site on asteroid Eros (*arrowed*). Two images from NEAR's descent, taken from 250 m and (*right*) 120 m, where transmission was interrupted as the probe landed. (NASA/JPL)

original mission objectives, the probe was commanded to approach the asteroid and make a slow 6 km per hour touchdown on its dusty, boulder-strewn surface. Eros doesn't have a strong gravitational pull, so the probe's descent was rather gentle. It was so gentle, in fact, that it actually survived the hard landing, and in an orientation that allowed communications to be re-established. Images taken by NEAR-Shoemaker during its fall show progressively more detail, and the last image, returned from an altitude of just 120 m, shows an area measuring 6 m across with boulders, rubble and loose soil (Fig. A.23).

Views of Deep Space

Launched in October 1998, NASA's Deep Space 1 probe was a test bed for ion propulsion and eleven other advanced technologies. The mission's science target was 9969 Braille, a Mars-crossing asteroid. The probe passed just 26 km from

Braille on July 29, 1999. Because of problems with the craft's tracking system no data was collected at the time of closest approach, and the two images and three infrared spectra that were secured during the encounter were taken from a distance of around 14,000 km. The low-resolution images of Braille showed it to be a tiny, irregular-shaped object measuring just $2 \times 1 \times 1$ km.

After its objectives were met in September 1999, NASA then extended the mission, taking advantage of the ion propulsion and other systems to undertake a challenging cometary encounter—a flyby of 19/PBorrelly, a comet that was discovered in December 1904. En route to the comet, just 2 months after completing its primary mission, the craft lost its navigational camera, but engineers managed to successfully re-establish contact.

Deep Space 1 successfully passed within 2,200 km of 19/P on September 22, 2001. The spacecraft was able to use all four of its instruments at Borrelly, enabling black-and-white images, infrared spectrometer measurements, ion and electron data, magnetic field and plasma measurements around the comet to be secured. The probe began taking black-and-white images of the comet 32 min before its closest pass to its nucleus. The most detailed image of the $8 \times 4 \times 4$-km nucleus was taken just a few minutes before closest approach. During the 2 min of closest approach, the camera was turned away so that the ion and electron instruments could examine Borrelly's inner coma.

Incredibly, analysis of the data revealed that no frozen water lies on the surface of Borrelly's nucleus. The comet is believed to have plenty of ice beneath its pitch-black surface, but any ice exposed to sunlight has vaporized away. Borrelly is unusually dark for an object in the inner Solar System, and its surface is possibly the darkest known in the entire Solar System—as dark as the print in this book (Fig. A.24).

Stardust's Fantastic Voyage

Launched on February 7, 1999, NASA's Stardust mission was to collect samples from the coma of 81P/Wild and return them to Earth. While en route, the probe's measurements of interstellar dust particles indicated the surprising presence of large tar-like molecules. The size of these molecules, with nuclear masses of up to 2,000 units (water's mass is 18 units), surprised scientists as much as the apparent absence of any mineral constituents, since only organic molecules can reach the size observed.

Stardust's Cometary and Interstellar Dust Analyzer (CIDA) found that five macromolecules hit it in 1999. Similar molecules could have played an important role in the development of life on Earth, constituting a delivery system for complex molecules believed essential for life to have started several billions of years ago. The particles might represent one means by which prebiotic molecules arrived on Earth. Other theories speculate that they came to Earth in asteroids or comets. Alternatively, they may have formed naturally on Earth itself. In any case, they are

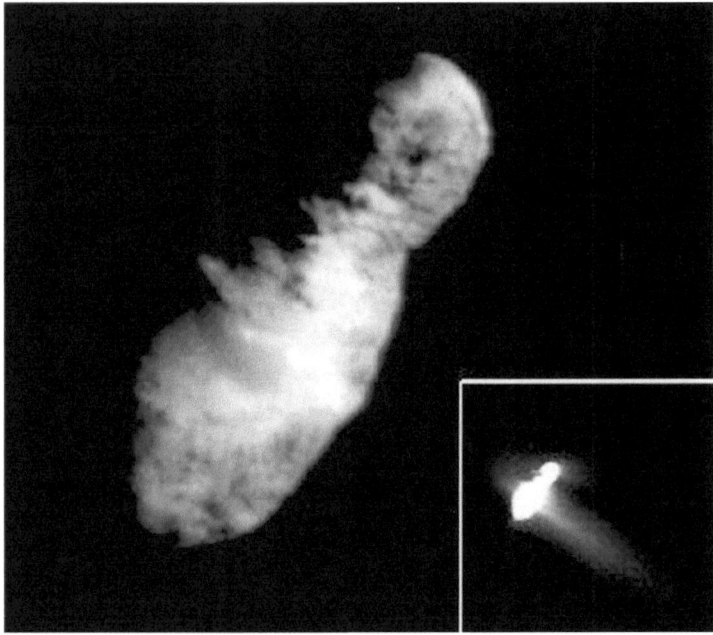

Fig. A.24 Astronomers were bowled over when they viewed this image of Comet Borrelly's 8-km-long bowling pin-shaped nucleus from a distance of 3,417 km. Smoother, brighter parts of the nucleus appear to be the source of the dust jets visible in the coma (inset). (NASA)

thought to be an essential ingredient of the primordial soup—that chemical cocktail from which life on Earth first sprang.

While en route to its encounter with 81P, it was decided to put Stardust through a trial run with flyby techniques when it passed asteroid 5535 Annefrank on November 2, 2002. Stardust passed the asteroid at a distance of 3,079 km and returned medium-resolution images showing a dark, irregular-shaped object measuring 7×5×3 km (Fig. A.25).

Stardust reached comet 81P/Wild on January 2, 2004, returning detailed images of its nucleus and capturing samples of its dusty coma for return to Earth. NASA's first dedicated sample return mission to a comet successfully flew through the dust and gas-laden coma surrounding the comet, passing within just 240 km of the nucleus. Active cometary nuclei are risky environments for spacecraft to fly past at high speed. Stardust was traveling at more than 6 km per second. The impact of a solid particle the size of a grape seed had the potential to disable the probe. Thankfully, its hefty impact shields served their purpose admirably, and despite taking a number of hits, Stardust emerged intact and fully functional.

Samples of the coma, captured in a special scoop during the encounter, were stored safely on board and returned to Earth in a capsule on January 15, 2006. It was to be the first chance that scientists ever had to analyze fresh cometary material (Fig. A.26).

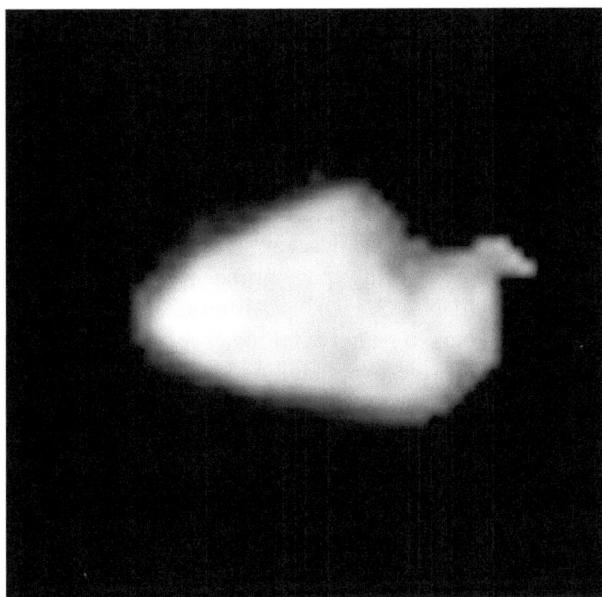

Fig. A.25 Asteroid Annefrank, imaged in November 2002 by the Stardust probe. (NASA/JPL)

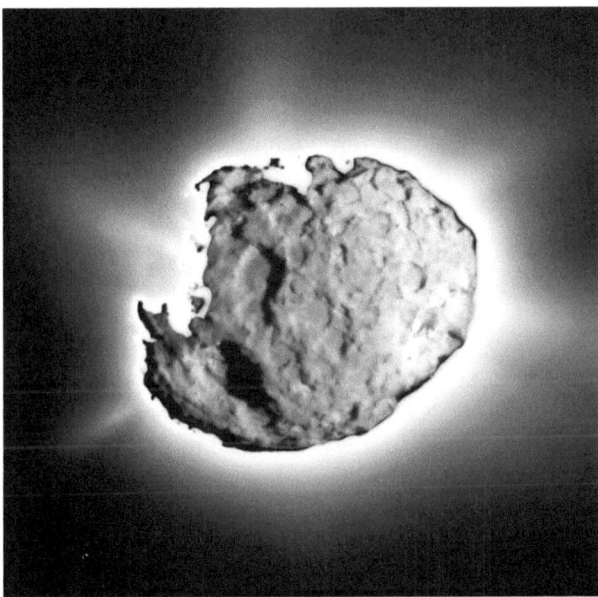

Fig. A.26 Stardust's camera captured stunning images of the comet 81P/Wild's rounded, highly pockmarked nucleus. This composite image, taken at the time of the probe's closest approach, shows the nucleus (short exposure) surrounded by its active jets of gas and dust (longer exposure). (NASA/JPL)

After completing its primary mission, the still healthy Stardust spacecraft was given a new mission. Renamed Stardust-NExT (New Exploration of Tempel 1), the spacecraft was to fly by 9P/Tempel—a comet that had already been visited and assaulted by the Deep Impact mission in 2005 (see below). Stardust-NExT flew past 9P/Tempel on February 14, 2011.

Having journeyed back to the vicinity of Earth, Stardust's sample return capsule separated and re-entered the atmosphere on January 16, 2006, streaking back at a velocity of 46,000 km per hour. The descent proved to be the swiftest return velocity ever experienced by a re-entering space probe. The re-entry was carefully monitored by scientists aboard NASA's DC-8 Airborne Laboratory (operated by the University of North Dakota) to gather vital data for studying spacecraft re-entries. Parachutes lowered the capsule to its landing spot at the Utah Test and Training Range, and after recovery it was taken to the Planetary Materials Curatorial facility at Johnson Space Center in Houston to be opened for analysis.

The precious material that had been collected by the spacecraft 2 years earlier consisted of minute grains of dust sampled from the coma of comet 81P/Wild. Once captured, the cometary material—embedded within a special low-density substance called an aerogel—was stowed within a secure compartment in Stardust's sample return capsule. It was the first time that pristine material had been directly sampled from the vicinity of a comet and returned to Earth for study. Also captured within the aerogel were interstellar particles—bits of matter originating from beyond the Solar System.

When the sample trays were first examined, hundreds of particles could be seen in the collector tray for cometary samples, each preceded by a narrow carrot-shaped track through the aerogel collecting medium. The largest track was almost big enough to put a little finger inside, while two large particles appeared to have exploded inside the aerogel after impacting on the tray's aluminum frame. The samples are thought to represent pristine material left over from the formation of the Solar System 4.6 billion years ago. It is estimated that more than a million specks of dust are embedded in the aerogel (Fig. A.27).

Deep Impact

Launched in January 2005, NASA'S Deep Impact spacecraft was on a mission that promised to pack quite a punch. It was intended to study the interior composition of Comet 9P/Tempel (Tempel 1) by releasing an impactor that would slam into the nucleus while the main craft recorded the event from a safe distance.

Released by its Deep Impact mother craft on July 3, 2005, the impactor probe sped towards the cometary nucleus—a potato-shaped lump of ice and rock some 14 km long—for 22 h. Traveling at a velocity of around 10 km per second (37,000 km per hour) the impactor probe transmitted a stream of images as it plunged towards its target. During its final few hours, topographic detail on the nucleus became ever clearer, as plateaus, hills, ridges and craters were resolved in increasing detail. The final images taken by the impactor showed features less than a meter across.

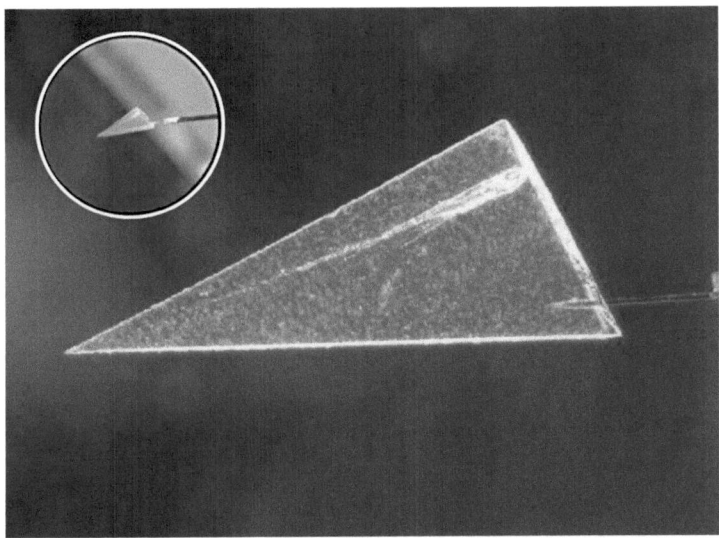

Fig. A.27 Scientists can extract and study particles embedded in aerogel in several ways. The image shows how a particle and its track have been cut out of the aerogel in a wedge-shaped slice called a keystone. A silicon 'pickle fork' is then used to remove the keystone from the remaining aerogel for further analysis. (NASA/JPL)

Impact occurred at 05:52 UT on July 4. The impactor penetrated the comet's crust and rapidly came to a halt; its kinetic energy was converted into an explosion equivalent to 4.5 metric tons of TNT. Immediately afterwards, the Deep Impact mother craft recorded a tremendous blast as material was thrown out by the explosion—first a brilliant flash, then a shock wave followed by a rapidly expanding plume of dust and ice that cast a shadow onto the nucleus. Nearing the nucleus, Deep Impact's mother craft went into shielded mode, but once beyond the comet its cameras looked back at the scene; the ejecta remained brilliant, and extended to several thousand kilometers.

Since the images of the resulting crater weren't as good as expected, the Stardust probe was redirected to 9P/Tempel to take follow-up images. Renamed Stardust-NExT (New Exploration of Tempel 1), the spacecraft passed 9P/Tempel on February 14, 2011. The crater that had been blasted out artificially more than 5 years previously was faint (Figs. A.28, A.29, A.30, A.31, and A.32).

Deep Impact Reloaded

After its encounter with 9P/Tempel, the Deep Impact mission was rebranded EPOXI—an amalgam of Deep Impact Extended Investigation (DIXI) and Extrasolar Planet Observation and Characterization (EPOCh). Comet 103P/Hartley was to be the target of the spacecraft, while EPOCh was to be conducted en route.

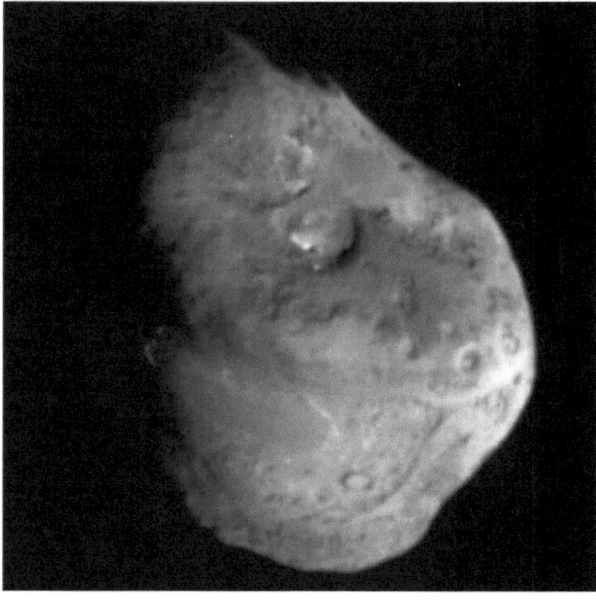

Fig. A.28 Comet 9P/Tempel imaged by Deep Impact's high-resolution camera. (NASA/JPL)

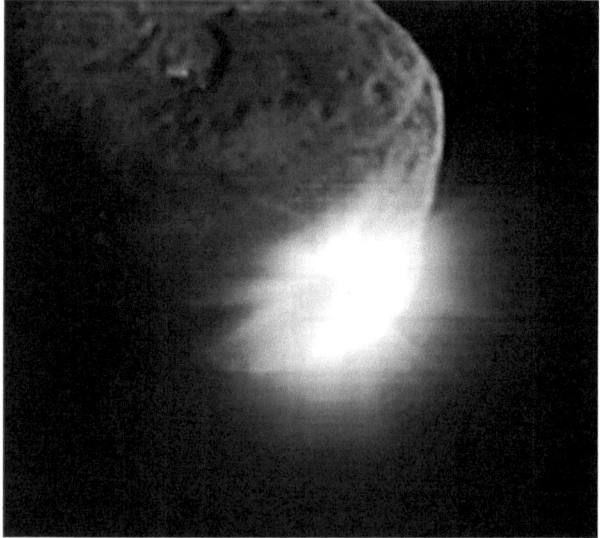

Fig. A.29 A bright plume of ejected material appeared above comet 9P/Tempel within moments of impact. Image by Deep Impact's high-resolution camera 13 s after impact. (NASA/JPL)

Fig. A.30 The Hubble Space Telescope imaged a marked brightening of comet 9P/Tempel (by about two magnitudes) immediately after the impact. (NASA/JPL)

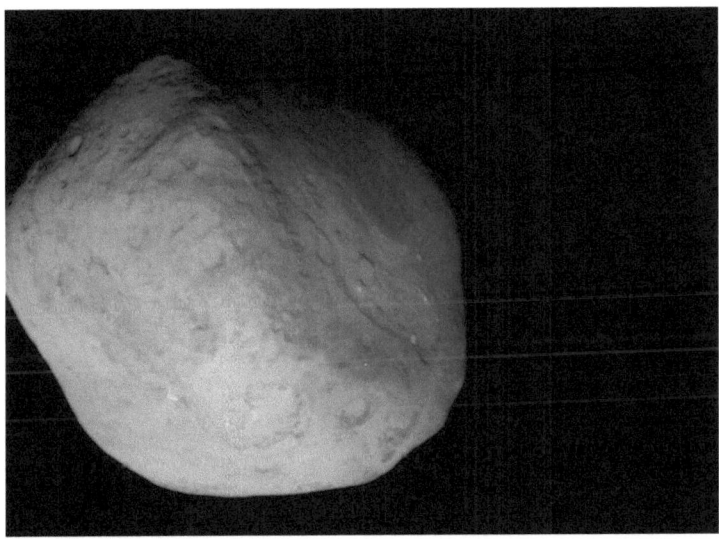

Fig. A.31 Comet 9P/Tempel imaged by Stardust-NExT in February 2011. (NASA/JPL)

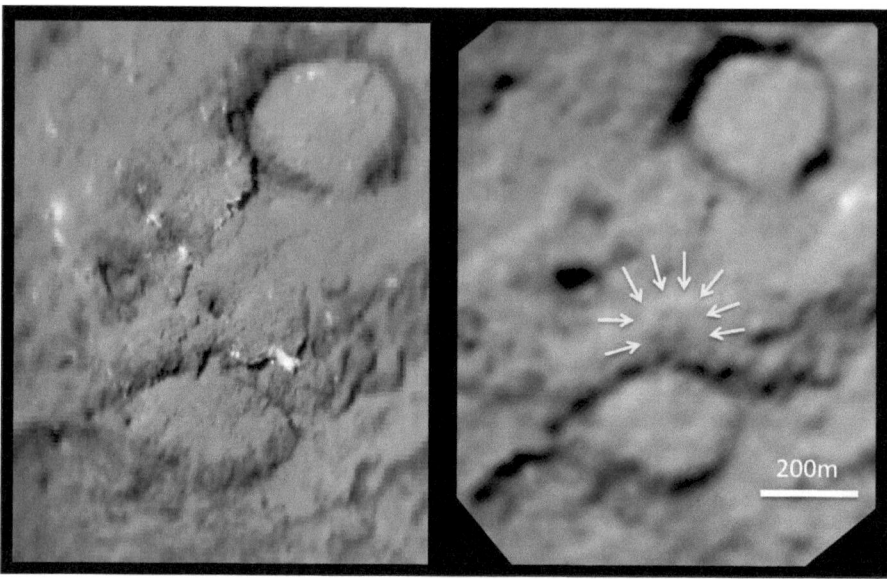

Fig. A.32 Comparison images of the impact site of the Deep Impact impactor—Deep Impact's image before impact in 2005 (*left*) and long after the impact, by Stardust in 2011 (*right*). (NASA/JPL)

103P/Hartley was enjoyed by amateur astronomers through binoculars and telescopes during October and November 2010 as it glided through Cassiopeia, Perseus and Auriga, almost reaching naked-eye visibility at its closest point to Earth on October 18. Adding yet more excitement to what was a very nice telescopic sight, the comet's active nucleus was imaged in glorious detail by the medium-resolution camera on board EPOXI as the spacecraft flew past the comet on November 4, 2010, from a distance of about 700 km. The comet's nucleus, or main body, is approximately 2 km long and 400 m at the 'waist,' its narrow central portion. Jets of sublimating gas mixed with dust were seen streaming out of the nucleus on the sunlit side (Figs. A.33, A.34, and A.35).

Hyabusa Hits the Mark

A project of the Japanese Aerospace Exploration Agency (JAXA), the spacecraft Hayabusa was launched from Kagoshima Space Center in 2003. Hayabusa's mission seemed doomed from the start, as solar flares affected its solar panels, reducing power for its ion drive and delaying arrival by several months. Its pointing control system failed as two out of three momentum wheels failed.

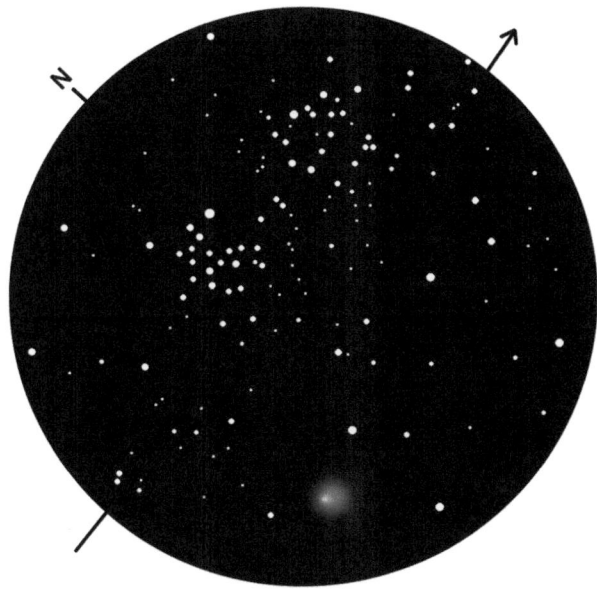

Fig. A.33 103P/Hartley makes a wonderful sight near the Double Cluster in Perseus, observed by the author at 20:40 on October 8, 2010, through 25×100 binoculars. (Illustration © by the author)

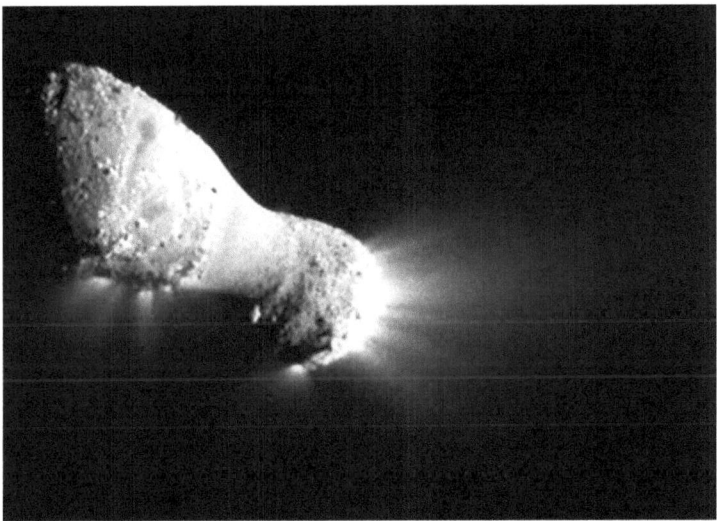

Fig. A.34 The active nucleus of 103P/Hartley, imaged by NASA's EPOXI spacecraft on November 4, 2010. (NASA)

Fig. A.35 103P/Hartley, observed by Paul Stephens with a 12-in. Newtonian (90×) on November 10, 2010, at 02 h UT. (Illustration © Paul Stephens. Used with permission)

Despite these problems, Hayabusa reached near Earth asteroid 25143 Itokawa on September 12, 2005, where it was 'parked' at a distance of a few kilometers. The probe's objectives were to study the asteroid with near-infrared and gamma ray spectroscopy and visible light imagery. A dust sample was to be obtained by firing a projectile at Itokawa. Images revealed that Itokawa is one of the Solar System's oddest-looking places—a loosely bound potato-shaped rubble pile some 500 m long.

Hayabusa carried a landing probe called Minerva to study the asteroid more closely. On its release from Hayabusa, the paint can-sized lander drifted off into space and was lost. In its attempts to collect a sample, Hayabusa made several efforts to approach Itokawa. It eventually succeeded on November 26, in an area dubbed 'Muses Sea' (a nod to the original project name MUSES-C), and the projectile was fired on the surface. A return home, with samples of Itokawa collected in the scoop, was put on hold for 3 years because of the need to address problems with the attitude control system, the power supply and radio.

Arrival back at Earth took place on June 13, 2010, when its precious asteroid samples were parachuted into the deserts of the south Australian outback while the spacecraft broke up and was incinerated in a large fireball. Around 100 particles were collected by the sample canister; mainly less than 10 mm in size, their composition was similar to meteorites and matched chemical maps of Itokawa obtained by Hayabusa from orbit. The dust is thought to have been exposed on the asteroid for about 8 million years (Fig. A.36).

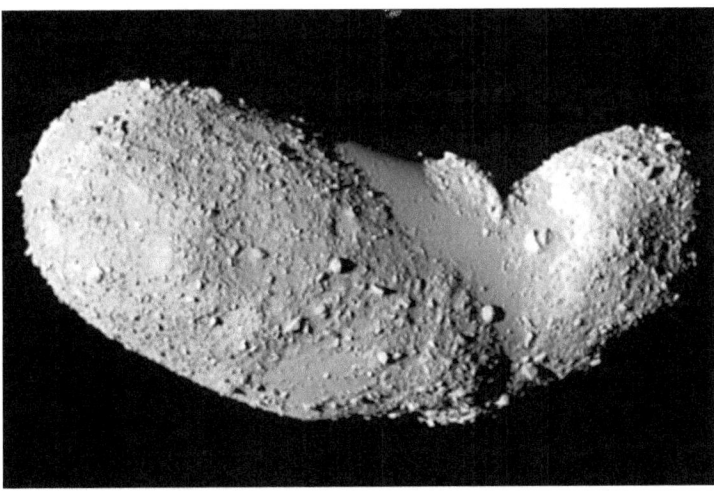

Fig. A.36 Asteroid Itokawa, imaged in 2005 by Japan's Hyabusa probe. (JAXA)

Rosetta: Two Asteroids and a Comet

The European Space Agency's space probe Rosetta, launched in March 2004 with a primary objective to study comet 67P/Churyumov-Gerasimenko, made flybys of two asteroids on its journey out. Asteroid 2867 Šteins was visited on September 5, 2008. Passing 800 km from the asteroid at a relatively slow relative velocity of 30,000 km per hour, Rosetta returned images of a diamond-shaped body measuring 7×6×4 km; a large crater some 2.1 km across, named Diamond, dominates the wider section of Šteins (Fig. A.37).

On July 10, 2010, the Rosetta flew by asteroid 21 Lutetia, passing it within 3,200 km at a relative velocity of 54,000 km per hour. Around half of the asteroid's surface was covered in 462 images, the most detailed ones at a resolution of 60 m per pixel. Rosetta also made observations with its visible/near-infrared imaging spectrometer, plus measurements of the magnetic field and plasma environment. Lutetia was found to be elongated, measuring 120 km on its major axis; heavily cratered, its largest impact feature, Massilia, is 45 km across. Grooves and scarps show deep fracturing, and the asteroid's high density suggests that the rock is rich in metal. Lutetia is likely to be a survivor from the Solar System's beginning (Fig. A.38).

Having entered deep-space hibernation mode in June 2011, Rosetta will be reawakened in January 2014 prior to its scheduled rendezvous with comet 67P/Churyumov-Gerasimenko. Approach to the comet takes place between January and May 2014, followed by a mapping phase in August and a landing in November, after which the comet will be escorted in its orbit until December 2015.

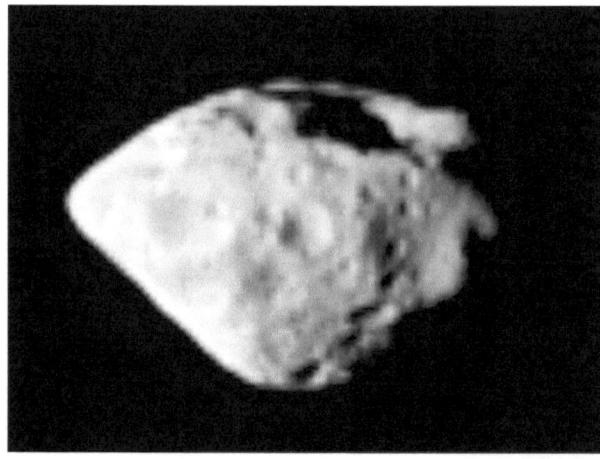

Fig. A.37 Asteroid Šteins, imaged in September 2008 by ESA's Rosetta probe. (ESA)

Fig. A.38 Asteroid Lutetia, imaged in July 2010 by ESA's Rosetta probe when at its closest approach. (ESA)

Dawn's Double Encounter

NASA's Dawn spacecraft was launched in September 2007 with the objectives of rendezvousing with and orbiting asteroid 4 Vesta for 9 months beginning in the summer of 2011, after which it would travel to dwarf planet Ceres to study it in detail, beginning in February 2015.

Vesta is a large rocky object, while Ceres is thought to contain large quantities of ice and volatiles. Dawn would cover the distance of 4.8 billion km in 8 years with the use of its advanced ion drive, which is ten times more efficient than conventional propulsion. Ion engines work by ionizing (giving an electric charge to) atoms of xenon gas; a high voltage charge shoots these ions through a metal grid at very high speed, around 144,000 km per hour, and the resulting reaction of the ions leaving the thrusters pushes Dawn in the other direction. Dawn's ion propulsion system does not produce quick bursts of speed, like a conventional rocket thruster. In fact, it would take Dawn 4 days to accelerate from zero to 100 km per hour. The thrusters would accumulate over 2,000 days of operations during the 8-year mission, allowing the thrust to gradually build over time to generate high speeds.

Dawn finally entered orbit around Vesta, the first of its mission targets, on July 16, 2011. Measuring $573 \times 557 \times 446$ km, it is the largest and most massive of the minor planets, second only in the main belt to dwarf planet Ceres. It is a differentiated object, with a crust, mantle and core. Its surface is rough and diverse, with light and dark markings. Some of its rocks resemble terrestrial volcanic rocks, rich in magnesium and iron, but it also has rocks melded by the heat of impact, as well as smooth 'ponds' of fine dust that were deposited in ejecta blankets. Vesta's south polar region has a huge impact crater, Rheasilivia, that boasts a bulging central peak some 22 km high. A massive collision gouged out 1 % of the asteroid's volume, blasting over 2 million cubic km of rock into space.

Some of the craters in the southern hemisphere, including Rheasilivia, are less than a couple of billion years old. It was confirmed that a group of meteorites identified on Earth, rich in iron and magnesium, actually came from Vesta—originally blasted into space as impact ejecta from these very craters. Indeed, Vesta is one of the largest sources of meteorites found on Earth. Around 6 % of all meteorites come from this asteroid (Figs. A.39 and A.40).

Dawn heaved itself out of orbit around Vesta on September 5, 2012, after having studied the asteroid for 14 months. It is set to reach the very different environment of dwarf planet Ceres in 2015. Big, primitive and icy, rounded and looking very much like a real planet, Ceres may have a thin atmosphere, and it could be host to hydrological processes leading to seasonal polar caps of water frost (Fig. A.41).

Fig. A.39 Asteroid Vesta, a mosaic of images captured by Dawn during its orbit in 2011–2012. (NASA/JPL)

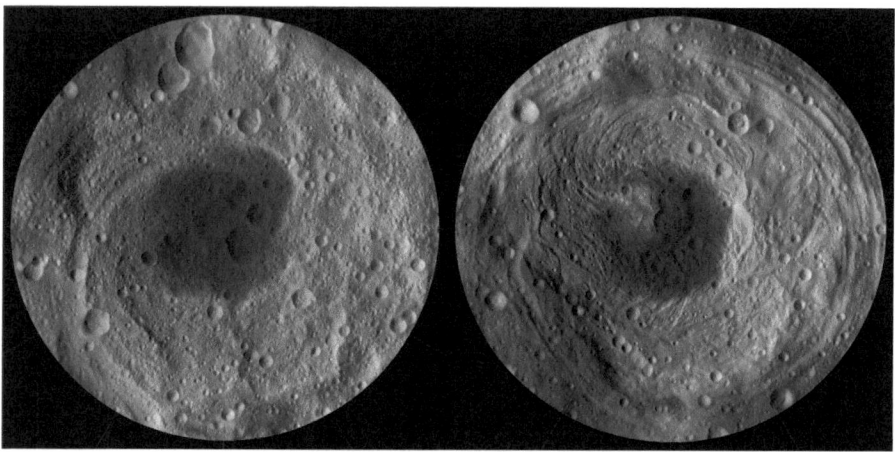

Fig. A.40 A color-coded topographical chart of Vesta's northern and southern hemispheres derived from Dawn's image analysis. Colors represent distance relative to the center of Vesta, with lows in violet/blue and highs in red. In the northern hemisphere the surface ranges from lows of −22 km to highs of 44 km, while lows of −38 km and highs of 43 km are found in the south. The crater Rheasilivia and its huge central uplift dominates the southern topography. (NASA/JPL)

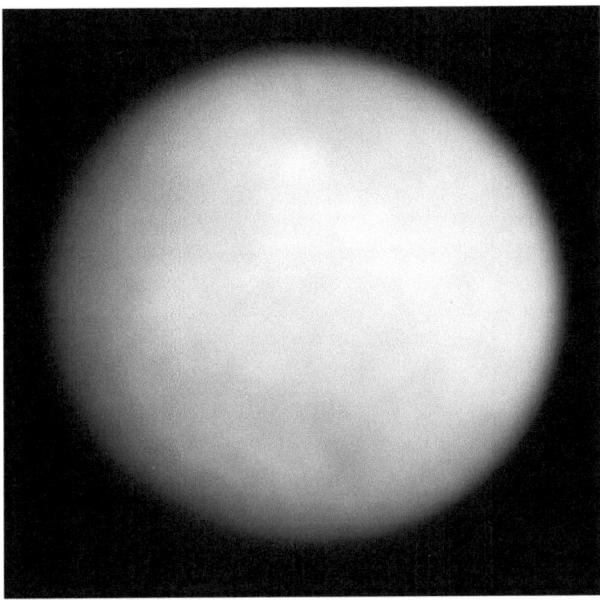

Fig. A.41 Ceres, imaged in January 2004 by the Hubble Space Telescope. (NASA)

Moon and Beyond with Chang'e 2

Launched in October 2010, China's Chang'e 2 space probe first paid a visit to the Moon. In lunar orbit between October 9, 2010, and June 8, 2011, the probe completed what was claimed to have been the highest resolution map of the Moon's surface. After leaving the Moon, Chang'e 2 proceeded to a gravitationally stable point in Earth's orbit (the L2 Lagrangian point) to test the tracking and control network. Following this, it was sent to make a flyby of near-Earth asteroid 4179 Toutatis. Earth-based radar imagery had long established that the asteroid is a bi-lobed object around 7 km across. During its flyby on December 13, 2012, in which it approached Toutatis to 3.2 km, Change'e 2 found that the 5×2 km asteroid's surface is cratered, the largest crater measuring around 1 km across. Toutatis may be a loosely bound rubble pile, formed from the coalescence of two main bodies (Fig. A.42).

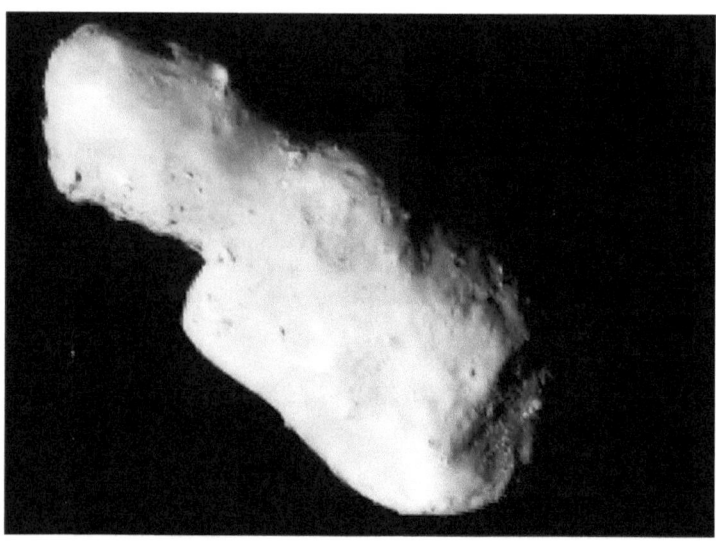

Fig. A.42 Asteroid Toutatis, imaged in December 2012 by the Change'e 2 space probe. (China Space Agency)

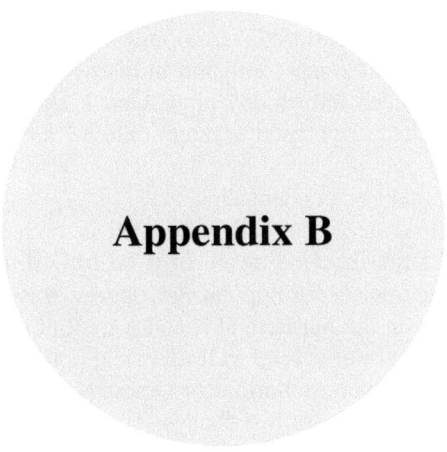

Appendix B

Some Recent Comets of Note: An Observer's Recollections

The late twentieth and early twenty-first centuries saw many comets make their way across the skies, but only a few of them were at all spectacular in terms of their apparent brightness. The news media are often guilty of hyping cometary apparitions—notably Halley in 1985–1986—a fact that can cause a great deal of disappointment among members of the general public who are led to expect much more of a visual spectacle.

What follows is a broad selection of some of the comets that have drifted across this author's field of view since 1983. It's by no means a comprehensive listing of all the comets I've set eyes upon, nor are they solely the brightest comets. I found it most enjoyable to go through my records and observations of years gone by and was surprised at how much I was able to recall of the excitement surrounding each comet and the pleasure that was to be had in sketching them at the eyepiece.

C/1983H1 (Iras-Araki-Alcock)

Having been an active amateur astronomer since 1980, no better object could have served as my own personal introduction to the world of comets than Iras-Araki-Alcock. It was bright enough to be seen from suburbia without optical aid; it was impressively large, came extremely close to Earth and moved at a blistering pace across the northern skies.

P. Grego, *Blazing a Ghostly Trail: ISON and Great Comets of the Past and Future*, 257
The Patrick Moore Practical Astronomy Series, DOI 10.1007/978-3-319-01775-4,
© Springer International Publishing Switzerland 2014

The comet was first seen as a 6th magnitude object on May 3, 1983, by veteran amateur comet hunter George Alcock at Peterborough, England (using his 15×80 binoculars from indoors, through a closed window) and the Japanese amateur Genichi Araki. The orbiting Infrared Astronomical Satellite (IRAS) had picked up the comet on April 25, but some confusion in distributing the news meant that it wasn't formally announced until a day after Alcock and Araki had discovered it visually. By May 8 it had brightened to magnitude 3.7 and was easily visible with the unaided eye; its coma measured less than a degree in width and contained an offset, almost stellar, nuclear condensation.

Comet Iras-Araki-Alcock was well publicized in both the general and astronomical media, and I first became aware of it on May 9 through a special *News Circular* issued by the Society for Popular Astronomy. Iras-Araki-Alcock was well placed for observation in the northern sky, and a spell of good weather conditions during the apparition made it a very well observed comet. It was a swift moving object, passing through the constellations of Draco, Ursa Minor, Ursa Major, Lynx, Cancer and Hydra in the space of just 5 days. The comet was visible to the naked eye with little difficulty, even from light-polluted urban areas. On May 9, through binoculars and telescopes it could be seen near Gamma Ursae Minoris, a diffuse elongated patch of light some $2°$ across, a fan-shaped coma with a tiny bright nuclear condensation offset towards the Sun. At high magnifications the motion of the comet against the starry background could actually be discerned in real time, a fact that astonished most astronomers previously used to heavenly sights that usually appear static. By the following night it had reached first magnitude and was close to Alpha Ursae Majoris (Dubhe).

Comet Iras-Araki-Alcock's motion was so rapid that it appeared near the open star cluster M44 (the Beehive) in Cancer on the evening of May 11. On this date the comet made an extremely close passage to Earth, coming as close as 0.031 au (4.6 million km) and shining at magnitude 2. This represented the closest cometary encounter since D/1770L1 (Lexell), whose perigee on July 1, 1770, was just 0.0151 au (2.3 million km) from Earth. No longer visible from the northern hemisphere, Comet Iras-Araki-Alcock's perihelion took place on May 21, when it approached the Sun at a distance of 0.99 au (148 million km) (Figs. B.1, B.2, and B.3).

Halley's Comet, 1985–1986

This most famous cometary visitor from the depths of interplanetary space was recovered in 1982 on CCD (a light-sensitive electronic device) images made with the almost equally famous 5-m reflector on Mount Palomar. When recovered, Halley's Comet was exceptionally faint and nearly 5 years away from becoming bright enough to be visible in average amateur telescopes. Like many other amateur astronomers in the northern hemisphere I followed Halley from early November 1985 to April 1986, with a break in early 1986 as the comet rounded the Sun at perihelion and was temporarily lost to view.

Fig. B.1 Comet Iras-Araki-Alcock, observed by the author through 12×50 binoculars on May 9, 1983, at 22:15 UT. (Illustration © by the author)

Fig. B.2 Comet Iras-Araki-Alcock, observed by the author through a 60-mm achromat (40×) on May 10, 1983, at 23:45 UT. (Illustration © by the author)

Fig. B.3 Comet Iras-Araki-Alcock, observed by the author through 12×50 binoculars as it passed near M44 on May 11, 1983, at 22 h UT. (Illustration © by the author)

An International Halley Watch was organized as a collection point for observations and to disseminate news about the comet. Professional astronomers lavished their full attention on the comet and employed the latest imaging techniques to squeeze as much information as possible out of Halley's precious light, and an armada of unmanned spacecraft constructed by Russia, Europe and Japan was sent to intercept the comet.

Halley's Comet entered the realm of amateur visual observation in October 1985, passing just half a degree south of M1 (the Crab Nebula) in Taurus; this was observed visually by some using large instruments, but it wasn't easy for most, owing to considerable skyglow caused by a full Moon in Aries. It became bright enough for regular 'backyard' observers in early November, when the waning Moon had moved a comfortable distance to the east; on November 5 it shone at the 7th magnitude and had a coma around 3 arcmin across.

I began searching on November 6, and observations showed that in the space of an hour the comet's motion among the stars was definitely perceived. As it passed by Kappa Tauri on November 11 it was easily visible telescopically as a diffuse patch of light, but no central condensation was obvious. On the following evening I noted the central condensation amid a coma that appeared bi-lobate, something resembling a trilobite's head. Some observers suspected a short tail.

On November 13 Halley's Comet appeared bright and asymmetric, with the nucleus situated towards the eastern part of the coma; the comet was easily visible through 7×50 binoculars. On November 15 the comet passed some 2° southeast of

Fig. B.4 Halley's Comet, observed and drawn by Paul Stephens a 6-in. Newtonian (30×) on November 12, 1985, at 00:45 UT. (Used with permission)

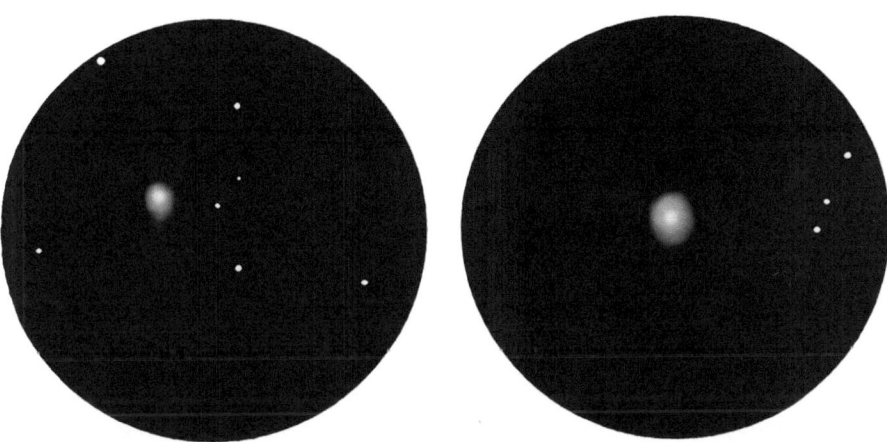

Fig. B.5 Halley's Comet, observed and drawn by the author through a 60-mm achromat (40 and 100×) on November 13, 1985, at 00:30 UT

the Pleiades, making a splendid sight through binoculars and telescopic wide fields; the coma was estimated to be around 7 arcmin in diameter. November 27 saw the comet nearest to Earth on its inbound leg into the inner Solar System—a perigee distance of 0.62 au (93 million km) (Figs. B.4, B.5, and B.6).

Fig. B.6 Halley's Comet and the Pleiades, observed and drawn by the author through 12×50 binoculars on November 17, 1985, at 00:35 UT

By December 3 Halley's Comet had moved into Pisces, south of the left hand side of the Square of Pegasus asterism. I served as a comet guide to passengers on two Halley's Comet special flights on the evenings of December 6 and 7. These chartered jets took a hundred or more eager comet-spotters from Birmingham Airport to a height of 6 km above the Irish Sea. With the aircraft banking to a suitable angle and the cabin blacked out, the comet was visible with the unaided eye, and an easy object to see through binoculars (which most of the passengers carried).

On December 18, a comparison made between the comet and the bright galaxy M31 in Andromeda (magnitude 4.5) showed that Halley was slightly dimmer. The first visual signs of a tail were discerned on December 28 when the comet was in Aquarius, just south of the celestial equator (Figs. B.7 and B.8).

On January 1, 1986, Halley's Comet crossed the Earth's orbit. In early January some observers enjoyed tracing the comet's tail for some 2° (using averted vision), but the lack of contrast caused by the city glow from my own location meant that it could only be definitely traced for around half a degree. Using out-of-focus comparisons, the comet was estimated to be magnitude 5.1 on January 3, just about visible with the unaided eye from a dark site. On January 11, my final observation prior to perihelion, the comet's short tail was visible without averted vision, and the coma was suspected by some visual observers to have a slightly greenish hue (Fig. B.9).

Fig. B.7 Halley's Comet, observed by the author through a 7-in. achromat (50×) on December 5, 1985, at 23:15 UT. (Illustration © by the author)

Fig. B.8 Halley's Comet, observed by the author through a 60-mm achromat (40×) on December 28, 1985, at 18:30 UT. (Illustration © by the author)

Fig. B.9 Halley's Comet, observed by the author through a 60-mm achromat (40×) on January 3, 1986, at 18:40 UT. (Illustration © by the author)

Heading into the glow of the evening twilight, the comet was lost to sight for a few weeks around its perihelion of 0.59 au (88 million km) on February 9, but it re-emerged into the morning skies and was favorably placed for observation from the southern hemisphere for the remainder of its apparition, although those in the northern hemisphere could follow it, low to the horizon, as it passed through Capricornus and Sagittarius in late February and early March, and again in late April and May as it passed through Hydra, Crater and Sextans. From the southern hemisphere the best views were to be had during late March and April, when Halley passed in front of the Milky Way in Corona Australis and Scorpius.

Halley's closest approach to Earth on its outbound leg took place on April 11, 1986, at a distance of 0.42 au (63 million km). I recovered Halley's Comet on April 28 when it was in southern Crater; using a 6-in. Newtonian I noted a half-degree long fan-shaped tail of two, possibly three components, with a detectable nuclear condensation. My last view of the comet was at 22 h UT on May 5, 1986, when I discerned a faint tail some 15 arcmin in length, with a suspected multiple tail; it was then around magnitude 6.5 and located in southern Sextans, in deep twilight just a hand's width above the southwestern horizon (Fig. B.10).

There was a tremendous amount of media hype surrounding Halley, even though astronomers had long predicted that the comet would not be a brilliant sight from the northern hemisphere. Some in the press blamed Halley when it became clear that

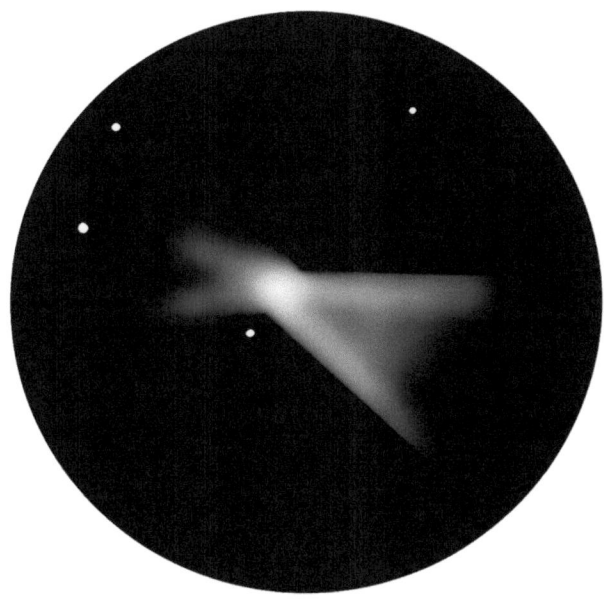

Fig. B.10 Halley's Comet, observed by the author through a 6-in. Newtonian (50×) on May 5, 1986, at 22 h UT. (Illustration © by the author)

the cometary visitor hadn't been reading its own publicity material! Typical newspaper reports accused Halley of being a 'crummy fuzzball' or a 'soggy squib.'

In February 1991, when the comet was outside Saturn's orbit, more than 2 billion km from the Sun and heading for a few decades of peaceful frozen semi-retirement in the outer Solar System, astronomers were astonished when Halley's nucleus flared to a hundred times its usual brightness. The first to spot this unprecedented outburst were astronomers Olivier Hainaut and Alain Smette using the European Southern Observatory's 1.54-m Danish telescope at La Silla in Chile. When the images were studied in detail it could be seen that a 300,000-km cloud of dust ejected by the nucleus was reflecting sunlight and was responsible for the increased brightness. Astronomers were baffled by this unexpected event. Models of cometary behavior had suggested that icy nuclei enter into a deep-frozen state beyond the orbit of Jupiter and are completely inactive at temperatures lower than −200 °C.

Halley's nucleus may have collided with a large meteoroid, and bits of Halley's crust at the impact site might have been blown off to form the observed temporary cloud of material. It is possible that the nucleus may have been severely fractured during this impact and weakened so much that it might fail to maintain its integrity during its next visit to the inner Solar System in 2061. It could disintegrate before our descendants' eyes.

C/1987 P1 (Bradfield)

Comet Bradfield, the brightest comet since Halley, was discovered as a magnitude 10 object by the veteran Australian comet observer Bill Bradfield on August 11, 1987. This, his 13th comet find, made him the most prolific discoverer of comets for a while (he was later outdone by the Shoemakers).

Comet Bradfield was a delightful little comet to observe, with a well-condensed head and a small thin tail, and it proved to be brighter than had originally been predicted. On October 10 it shone at magnitude 6.7, and by the 28th it had become just visible with the unaided eye from a dark site. It reached its brightest in mid-November when, located in Serpens, it shone at magnitude 5.4. Northern hemisphere observers were delighted as, during the space of 6 weeks between mid-November 1987 and early January 1988, it passed through the constellations of Aquila, Sagitta, and Delphinus and then into Pegasus, moving slowly northwards in declination. Fading below naked-eye visibility in mid-December, by the time it had reached Andromeda, near Alpha Andromedae on January 9, it shone at magnitude 7.5 (Fig. B.11).

Fig. B.11 C/1987 P1 (Bradfield), observed by the author through 11×80 binoculars on December 11, 1987, at 18 h UT. (Illustration © by the author)

23P/Brorsen-Metcalf

This periodic comet with an orbital period of 70 years has had three confirmed apparitions. It was first observed on the morning of August 11, 1847, by Theodor Brorsen who was conducting a routine comet sweep at Hamburg's Altona observatory. Its next return took place in 1919 when it was recovered on the morning of August 21 by Joel Metcalf at Camp Idlewild, Vermont, in the United States, confirming the comet's periodicity. As a result, the comet is named after the two astronomers. Comet Brorsen-Metcalf became a naked-eye object with a large coma, almost the apparent diameter of the Moon, during its 1919 apparition.

On July 4, 1989, the comet was recovered by Eleanor Helin using the 4.6-m Schmidt telescope at Mount Palomar. It was a considerable distance from the location predicted, which was based upon visual astrometric measurements made 70 years previously. From mid-July to late September Comet Brorsen-Metcalf carved a path across the northern sky, from Pisces, and through Traingulum, Perseus, Auriga, Lynx, Cancer and Leo. On July 30 it shone at magnitude 6.8, with a diffuse, 10-arcmin-diameter coma.

My own observations of the comet were made in the pre-dawn skies of August, several weeks before perihelion, and at this stage the head of the comet was just visible to the naked eye. Using a 150-mm rich-field reflector at a low power, the comet displayed a beautiful condensed nuclear region and sprouted a lovely 'spring onion' tail. Many observers took advantage of the total lunar eclipse on the morning of August 17 as the perfect time to view the comet. For a space of an hour or so the glare of full Moon was quelled, enabling the comet to be seen in all its glory, when it shone at its peak magnitude of 5. By September 5 the comet's narrow tail was some 6° in length and slightly distorted at a distance from the nucleus. Perihelion took place on September 11 at 0.48 au (72 million km) (Fig. B.12).

C/1990K1 (Levy)

On May 20, 1990, visually observing with a 4-m reflector, American astronomer David Levy discovered a tiny 10th magnitude comet in Andromeda. It was Levy's 6th comet discovery. As the comet traversed the familiar asterism of the Square of Pegasus, it brightened so that by June it was 9th magnitude, with a small rounded coma and a degree-long tail, giving it the appearance of a Spanish onion (rather than the spring onion of Comet Brorsen-Metcalf).

By August, Comet Levy was easily visible with the naked eye, and it made an impressive sight through binoculars. By mid-August the tail was some 1½° long and it shone at magnitude 4.3. A small outburst took place on August 16–17, with a surge in brightness and changes in the tail noted. Particularly noteworthy was the night of August 18 when the comet passed just a few minutes of arc away from the globular cluster M15 in Pegasus.

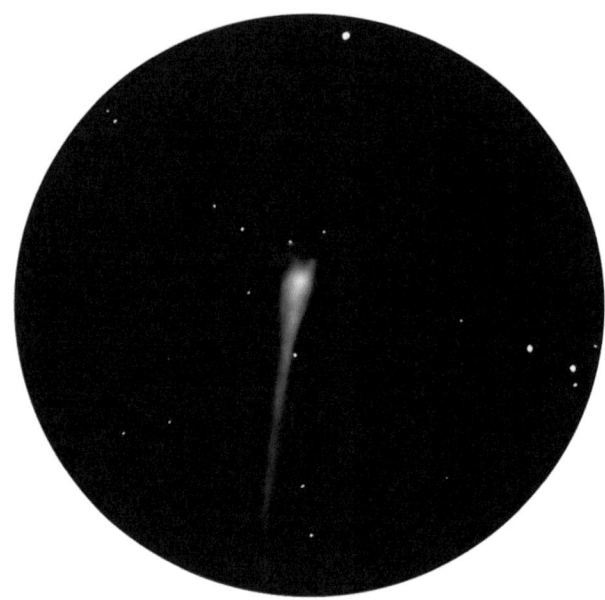

Fig. B.12 Comet Brorsen-Metcalf, observed by the author through a 6-in. Newtonian (30×) on August 28, 1989, at 03:30 UT. (Illustration © by the author)

On August 25th I observed Comet Levy move an angular distance of around 10 min of arc against the background star field, a real-time phenomenon of motion that I had only previously noticed with Comet Iras-Araki-Alcock over 7 years before. The comet was at its closest to Earth at the end of August, 0.43 au (64 million km) distant, and was visible from northern temperate latitudes until early September, when it shone at magnitude 4 (Fig. B.13).

C/1991T2 (Shoemaker-Levy)

American astronomers Gene Shoemaker and David Levy discovered this comet on October 6, 1991, when it was a tiny 16th magnitude dot on a photographic plate. During May to July 1992 the comet tracked through the northern skies from Cassiopeia to Ursa Major but never brightened enough to be visible with the naked eye. Through binoculars and telescopes Shoemaker-Levy appeared as a small fan-shaped smudge of light with a tiny star-like nuclear condensation—not the most spectacular of comets, but very easy to locate from the northern hemisphere. My own brightness estimates, made between July 6 and 19, when it crossed the bowl of the Plough asterism in Ursa Major, were from 7.5 to 7.9 (Fig. B.14).

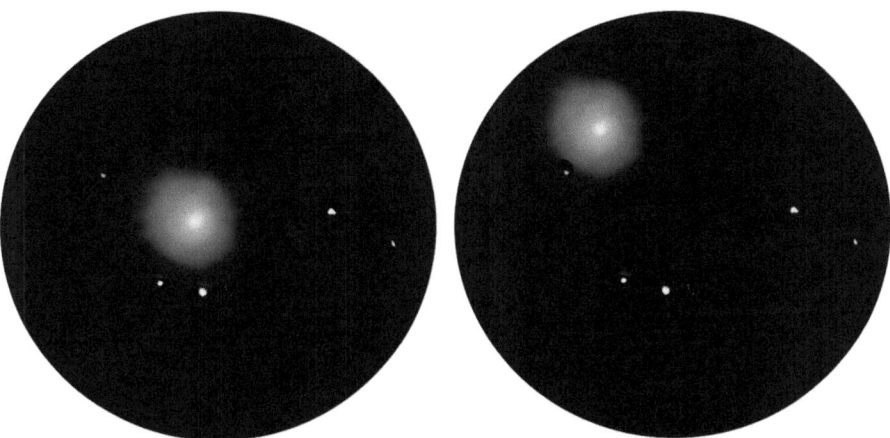

Fig. B.13 C/1990K1 (Levy), observed by the author through a 12-in. Newtonian (75×) on August 25, 1990, at 23:30 (*left*) and 22:45 UT (*right*), showing the comet's motion. (Illustrations © by the author)

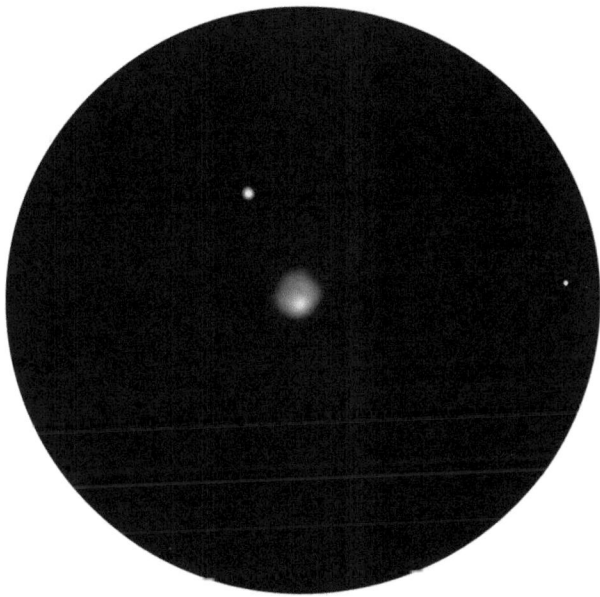

Fig. B.14 C/1991T2 (Shoemaker-Levy), observed by the author through 15×50 binoculars on July 6, 1992, at 22:45 UT. (Illustration © by the author)

109P/Swift-Tuttle

Comet Swift-Tuttle, the parent of the Perseid meteor stream, was recovered on September 26, 1992, by the Japanese amateur astronomer Tsuruhiko Kiuchi, who used a pair of giant 25×100 binoculars to spot Swift-Tuttle, then a faint 10th magnitude object in Ursa Major.

Kiuchi was at first unaware that this was anything other than a new comet. Most astronomers expected Swift-Tuttle to make its return appearance between 1979 and 1982, in accordance with the predicted 120-year orbital period. But this had been erroneously based on an identification of a comet observed in 1790 by Pierre Méchain (now called 8P/1790 A2) as the same object as Swift-Tuttle. But astronomer Brian Marsden reckoned that a comet observed by Ignatius Kegler from Beijing, China, in 1737 was the true past manifestation of Swift-Tuttle, and on that basis he calculated an orbit that brought the comet back to our vicinity in 1992. Marsden was proven right, and incredibly one of his two predicted dates of perihelion passage (December 11) was just a day out. Such a success has prompted serious suggestions that Swift-Tuttle should be renamed 'Marsden's Comet.'

As it passed through Ursa Major and Boötes during the course of October the comet brightened from magnitude 8.5 to 6, in mid-month displaying a tail one degree long. It had brightened to naked-eye visibility by early November, and by the middle of the month binoculars revealed a tail some $2°$ long and the comet shone at magnitude 5. It remained at about the same brightness into December, reaching its maximum brightness as it passed through Aquila. On December 3, observing through 12×50 binoculars, I estimated the coma as being the same apparent diameter as the Moon's Mare Crisium—about 3 arcmin across and the tail about half a degree long (the Moon was in its waxing gibbous phase at the time, so moonlight reduced these observed values).

Perihelion took place on December 10, following which it continued its path into more southerly declinations while its angular distance from the Sun decreased. Swift-Tuttle reappeared in southern hemisphere morning skies during April 1993, but its magnitude had faded to 10; moving through Grus, Tucana and Hydrus, the comet was set against the Large Magellanic Cloud by the beginning of June (Fig. B.15).

153P/Ikeya-Zhang

Kaoru Ikeya (discoverer of five comets in the 1960s) in Japan and Daqing Zhang in China—both amateur astronomers—discovered a new comet on February 1, 2002. It turned out that the comet, initially named C/2002 C1 (Ikeya-Zhang), was none other than the comet that had been observed 341 years earlier by Johannes Hevelius in 1661.

Heading towards the Sun, the comet reached perihelion on March 18, at 0.5 au (75 million km) from the Sun. Moving north through Cetus and Pisces into Andromeda, the comet passed a few degrees from the galaxies M31 (in Andromeda)

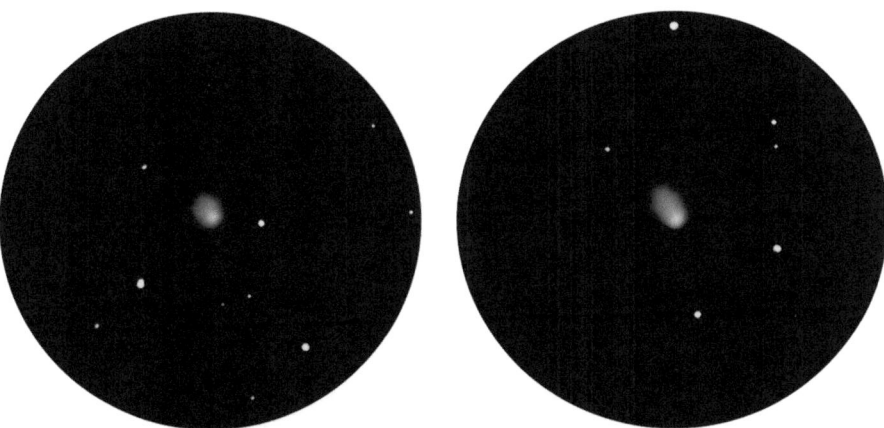

Fig. B.15 Comet Swift-Tuttle, observed by the author through 12×50 binoculars, showing its movement between November 16 (*left*) and 20 (*right*), 1992, both at 17:45 UT. (Illustrations © by the author)

and M33 (in Triangulum). Northern hemisphere observers were treated to the brightest comet since Hale-Bopp 5 years ago. Living up to some of the more optimistic predictions, Comet Ikeya-Zhang—officially designated C/2002 C1—reached magnitude 3.3 in late March as it climbed northwards through Pisces into Andromeda. Ikeya-Zhang will fade during May, from a predicted magnitude of 5.4 on May 1 to 8.1 on May 31. The comet passes just 2° north of M13 in Hercules on the evening of May 16 (Figs. B.16 and B.17).

C/2006 P1 (McNaught): The Great Comet of 2007

Rob McNaught's 31st comet discovery was made on August 7, 2006, using the 0.5-m Uppsala Schmidt camera at Siding Spring in Australia. A faint 17th magnitude at discovery, calculations of the orbit strongly suggested that it could become very bright when it reached perihelion at 0.17 au (25 million km) on January 12, 2007, providing it survived its passage so close to the Sun. Moreover, the comet was to pass between Earth and the Sun at around this time, producing a forward scattering of light that promised to increase its magnitude ten times (two magnitudes).

Coming within visual telescopic range in September, by November Comet McNaught had reached 9th magnitude, but was only visible from southerly latitudes. Observations revealed that the comet was significantly brighter than predicted; on December 29 it shone at magnitude 3.9, but in the first week of January it had reached 2nd magnitude. At this time there was a short window of opportunity for observers in north temperate latitudes to view the comet low in the twilight sky. I managed to observe the comet early on the evening of January 10, 2007. It was a spectacular sight in the twilight, with a tail several degrees long.

Fig. B.16 Comet Ikeya-Zhang, observed by the author through a 10-in. Newtonian on May 3, 2002, at 01 h UT. (Illustration © by the author)

Fig. B.17 Comet Ikeya-Zhang in close apparent proximity to globular cluster M13 in Hercules, observed by the author through 7×50 binoculars on May 16, 2002, at 01:50 UT. (Illustration © by the author)

Fig. B.18 The Great Comet of 2007, viewed by the author on the early evening of January 10, 2007, through 15×70 binoculars. (Illustration © by the author)

Photometric studies in daylight taken by Richard Miles of the British Astronomical Association on January 14 enabled him to determine a magnitude of −5 for the nuclear region. On the same day I managed to see the comet as a star-like point through binoculars during the daytime (with the Sun, around 6° to its west, conveniently shielded by a nearby roof).

From the southern hemisphere the comet was, of course, a truly Great Comet. As it moved south and its angular separation from the Sun increased it retained a bright, narrow head and developed a superb curving, filamentary dust tail that was even visible from the northern hemisphere, peeking up like multiple rays over the horizon while the head remained below. By January 18 the head, plus the tail and its striations, were visible with the naked eye from the southern hemisphere, the tail stretching from around 20–30°. During March Comet McNaught faded below naked-eye visibility, falling from the 6th to 8th magnitude, and the tail had shrunk to a degree or two in length (Figs. B.18 and B.19).

17P/Holmes

Discovered by Edwin Holmes from London on November 6, 1892, the comet experienced a couple of flare-ups, bringing it within naked-eye range. During its eight subsequent returns, Comet Holmes was a pretty faint and unspectacular object.

Fig. B.19 The Great Comet of 2007, viewed in daylight through 7×50 binoculars by the author on January 14, 2007. (Illustration © by the author)

But on October 24, 2007, the comet (5 months past opposition) underwent another outburst, suddenly brightening from magnitude 17 to magnitude 2.5—a rise amounting to about half a million times what it should have been. This was the greatest outburst ever observed in a comet, and its coma temporarily expanded to become the largest object in the Solar System, with a diameter greater than that of the Sun. As the comet approached opposition with Earth in late October its tail was pointing directly away from our point of view. The comet appeared to the naked eye as a large fuzzy disk about the same apparent angular diameter as the Moon.

Arcing south through Perseus, Comet Holmes' coma had grown to about a degree across by the end of November. At its largest around January 12, 2008, it was 2° wide; the coma displayed a distinct bow-shock, inside of which was a bright, offset nuclear region and a narrow tail. The comet faded considerably during February, falling below naked-eye visibility by the middle of the month. By mid-March, its nucleus was virtually undetectable telescopically, and its coma was exceedingly faint (Fig. B.20).

Fig. B.20 Comet Holmes, observed by the author on October 30, 2007, at 02:30 UT, using a 200-mm SCT (75×). (Illustration © by the author)

C/2011L4 (PanSTARRS)

The comet's peculiar name is derived from the instrument that discovered it back in June 2011: a 1.8-m telescope, part of the Panoramic Survey Telescope and Rapid Response System (Pan-STARRS) atop Haleakala on Maui, Hawaii. C/2011L4 arrived fresh from the Oort Cloud and experienced its parent star up close for the first time in 2013. It came closest to Earth, 1.1 au (165 million km) on March 5, 2013, and reached perihelion of 0.3 au (45 million km) 4 days later.

C/2011L4 was visible with the unaided eye to northern hemisphere observers between mid-March and early April. The comet, fading in brightness as it moved north, passed through Cetus and then Pisces until March 22, then proceeded through southern Andromeda until April 9, when it entered Cassiopeia, where it remained throughout the rest of April. From around March 26 onwards C/2011L4 required binoculars to be seen, and during early April it became circumpolar and visible above the horizon all night long.

By March 7 the comet, located in western Cetus and 16° east of the Sun, had just about nudged into the evening sky (3° high at sunset), but shining at magnitude 3.6 it was practically unobservable in the twilight without optical aid. At perihelion on the 9th C/2011L4 was 7° high at sunset—still a relatively elusive twilight object— but, given a clear western horizon, binoculars revealed its bright nucleus, coma and

Fig. B.21 C/2011L4, observed by the author on March 30, 2013, at 21:15 UT, using 25×100 binoculars. (Illustration © by the author)

tail. On March 13 the thin lunar crescent lay about 7° northeast of the comet, almost in line with the comet's tail; both objects were visible in a 7×50 binocular field of view, providing a rare and beautiful sight.

By March 18 C/2011L4, having faded to magnitude +4.4, had climbed sufficiently north for it to be visible above the northwestern horizon by the time astronomical night began, but, a high, ever brightening waxing Moon hampered the comet's naked-eye visibility.

C/2011L4 wasn't the naked eye spectacular that had originally been hoped for, but it was pretty good to view through binoculars or telescopes as it headed north. By the evening of March 30, the comet (magnitude +6.7) moved to within 6° of M31, the Andromeda Galaxy and closing, making another wonderful binocular sight—both objects could be encompassed in a 7 × 50 binocular field, and they appeared to be of about the same brightness, magnitude 4.3. M31 and C/2011L4 were at their closest on April 4–5, separated by just 2°. During April the comet passed numerous deep-sky objects in Cassiopeia, including less than a degree west of open cluster NGC 103 (magnitude +9.8) on April 24, the comet outshining it at magnitude 8.

A northern circumpolar object, the comet raced off into the distance through the spring and summer of 2013, passing through Cepheus, Draco and Ursa Minor, fading to 12th magnitude by July (Fig. B.21 and B.22).

Fig. B.22 C/2011L4, observed and drawn by Paul Stephens on April 1, 2013, at 20:40 UT, using a 7-in. Maksutov (60×). (Used with permission)

Glossary

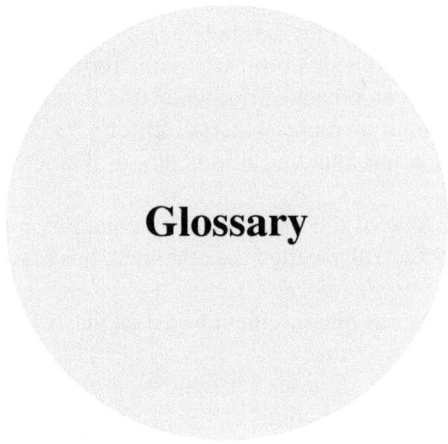

Achromat A refracting telescope whose objective lens is designed to minimize the effects of chromatic aberration, bringing different visual wavelengths of light to a near common focus. The devices are usually doublets consisting of two closely spaced lenses of different glass types and figured to a suitable shape. An apochromat telescope uses special glass with refractive properties that enables visible light to be focused to a near-perfect point; apochromats are much more expensive than achromats.

Accretion The gathering together of material. On a small scale, accretion works when two tiny particles collide and stick together. On a large scale, the gravity of individual particles attracts other particles, producing gradual growth of particle size. Accretion formed the planets and their satellites plus asteroids and comets.

Albedo A measure of an object's reflectivity. A pure white reflecting surface has an albedo of 1.0 (100 %). A pitch black, non-reflecting surface has an albedo of 0.0.

Altitude The angle of an object directly above the observer's horizon. An object on the horizon has an altitude of 0°, while at the zenith its altitude is 90°.

Aperture The diameter of a telescope's objective lens or primary mirror, usually measured in inches, millimeters, centimeters or meters.

Aphelion The point in an object's orbit furthest from the Sun.

Apochromat A refracting telescope whose objective lens is designed to minimize the effects of chromatic aberration, bringing different visual wavelengths of light to a near common focus. Apochromats usually have doublets consisting of two closely spaced lenses of different glass types figured to a suitable shape.

Apparition The period of time during which a planet, asteroid or comet can be observed between conjunctions with the Sun.

P. Grego, *Blazing a Ghostly Trail: ISON and Great Comets of the Past and Future*,
The Patrick Moore Practical Astronomy Series, DOI 10.1007/978-3-319-01775-4,
© Springer International Publishing Switzerland 2014

Arcminute   A measure of angle equal to 1/60 of a degree, often indicated with the symbol'.

Arcsecond   A measure of angle equal to 1/60 of an arcminute, or 1/3,600 of a degree, often indicated with the symbol ".

Apogee   The point in an object's orbit when it is furthest from Earth.

Aphelion   The point in an object's orbit when it is furthest from the Sun.

Asteroid   A large chunk of rocky material orbiting the Sun, ranging from a few tens of a meter to a few hundred kilometers in diameter. Also called a minor planet.

Asteroid Belt   The zone of the Solar System containing a large number of asteroids. The Asteroid Belt, also called the main belt, lies between the orbits of Mars and Jupiter.

Astronomical unit (au)   A measurement based on the average distance from Earth to the Sun—149,597,870 km.

Atmosphere   The mixture of gases surrounding a planet, satellite or star. A comet's atmosphere is called its coma.

Axis   An imaginary line around which a planet rotates.

Azimuth   The angular distance measured clockwise from north around the observer's horizon, given in degrees. North has an azimuth of 0°, east is 90°, south 180°, and west 270°.

Binoculars   An optical instrument consisting of a pair of small parallel refracting telescopes, allowing both eyes to view simultaneously. 10×50 binoculars have a magnification of 10 and lenses of 50 mm in diameter. Binoculars are an indispensable tool in the comet observer's toolkit.

CCD   Charge-coupled device; a light-sensitive electronic chip used in astrophotography.

Celestial sphere   From our perspective on Earth's surface, the stars appear attached to the inside of a vast, all-encompassing sphere. Earth's poles point directly to the celestial poles, and the celestial equator is in the same plane as Earth's equator. The celestial sphere appears to rotate around us from east to west as our planet spins on its axis.

Coma   The shell of dust and/or gas surrounding a comet's nucleus that forms a temporary cometary atmosphere, caused by solar heating and sublimation of icy volatiles. The released gas often jets out from the nucleus, taking with it dust grains; the pressure of sunlight sweeps the material away from the coma, forming a tail.

Comet   A large chunk of ice and rock that heats up on entering the inner Solar System, emitting gas and dust that forms a coma, and perhaps developing one or more tails of gas and dust.

Conjunction   Two or more close Solar System objects sharing the same right ascension (RA).

Constellation   A precisely defined region of the sky created to enhance our familiarity with the heavens. There are 88 recognized constellations, some of which date back to remote antiquity.

Core   The central region of a star or large planet—usually very hot and under extreme pressure. Many minor planets are chips from larger objects and may not have a well-defined core.

Crater   A feature, often depressed beneath its surroundings, bounded by a circular (or near-circular) wall. Almost all the large craters visible in the Solar System have been formed by asteroidal impact, but a few smaller craters are endogenic, which means of volcanic or explosive origin.

Culmination   The passage of a celestial object across the observer's meridian, when it is at its highest above the horizon.

Declination   On the celestial sphere, declination is equivalent to latitude on Earth. It is measured north (denoted by +) or south (−) of the celestial equator in degrees, minutes, and seconds of arc. The celestial equator is at a declination of 0°, while the north and south celestial poles are at +90 and −90°, respectively. Declination is used in conjunction with right ascension (RA) to pinpoint an object's location in the sky at any specified time.

Degree   1/360 of a circle. The Sun and Moon are about half a degree across. In terms of heat, degrees are increments of a temperature scale. Scales most commonly used in astronomy are Celsius (C) and Kelvin (K). 0 °C, or 273.16 K, is the freezing point of water. 0 K, or −273.16 °C, is known as absolute zero, where all molecular movement ceases.

Eccentricity   A measure of how an object's orbit deviates from circular. A circular orbit has zero eccentricity. Eccentricity between 0 and 1 represents an elliptical orbit.

Ecliptic   The line traced out on the celestial sphere by the Sun throughout the year, corresponding with the plane of Earth's orbit around the Sun. The Moon, planets and most minor planets orbit in planes roughly the same as the ecliptic.

Electromagnetic radiation   All energy in the electromagnetic spectrum—from short wavelength gamma rays to long wavelength radio waves—is propagated through space at the speed of light by vibrating electrical and magnetic disturbances. Visible light is a form of electromagnetic radiation.

Elongation   The apparent angular distance of an object from the Sun, measured between 0 and 180° east or west of the Sun.

Ephemeris   A table of numerical data or graphs that gives information about a celestial body, such as its declination and right ascension, in a date-ordered sequence.

Equator   The great circle of a celestial body whose plane passes through its center and lies perpendicular to its axis of rotation.

Eyepiece   A lens inserted into a telescope that serves to magnify the image formed by the objective lens or primary mirror and focus it into the eye.

Fireball   A very bright meteor caused by a large meteoroid's passage through Earth's atmosphere.

Gas giant   A very large planet composed largely of gases—mainly hydrogen and helium. Jupiter, Saturn, Uranus and Neptune are the Solar System's gas giants. They have no solid surface.

Geocentric   Having Earth at its center. The geocentric theory of the universe postulated that everything in the cosmos revolved around our planet.

Heliocentric   Having the Sun at its center. The heliocentric theory postulated that Earth and the planets revolved around the Sun.

Impact crater   An explosive excavation formed by a large projectile striking at high speed.

Kuiper Belt   A region of the outer Solar System beyond the orbit of Neptune to about 50 au from the Sun that probably contains many tens of thousands of icy planetesimals (cometary nuclei known as Kuiper Belt Objects—KBOs) larger than 100 km in diameter. More than a thousand KBOs are now known. Named after Gerard Kuiper, who proposed its existence in the mid-twentieth century.

Light pollution   Any unnecessary emission of light into the night sky from artificial sources, such as residential, commercial or industrial lighting. Small scale local light pollution, such as a glaring streetlight overlooking an observing site, affects the individual's ability to adapt to the dark, since the pupils cannot dilate fully when exposed to direct light. Large-scale light pollution reflects from particles in the atmosphere, causing widespread skyglow over urban and industrial areas, reducing the contrast between faint diffuse celestial objects and the dark sky and making them difficult or impossible to see.

Light year   The distance traveled by light in 1 year. At a velocity of 300,000 km per second, light travels around 10 trillion km in a year.

Limb   The apparent edge of a planet or natural satellite.

Magnitude   The perceived brightness of a celestial object is called its *apparent magnitude*. A jump from one magnitude to another corresponds to a brightness change of about 2.5 times. The brighter the object, the lower the value, i.e., an object of magnitude 1 is 2.5 times brighter than one of magnitude 2. The difference between a magnitude 0 object and one of magnitude 5 is a brightness difference of 100 times. Objects brighter than magnitude 0 are given a negative magnitude. The brightest star, Sirius, is magnitude −2.5, while the faintest stars visible with the unaided eye from beneath a dark sky are around magnitude 6. A star's real brightness, called *absolute magnitude*, takes into account the object's distance.

Meteor   A flash of light caused when a meteoroid burns up on entering Earth's (or another planet's) upper atmosphere.

Meteoroid   A small lump of rock or metal in space. If it survives all the way down to Earth's surface it is called a *meteorite*.

Minor planet   See asteroid.

Moon   The Moon is Earth's only natural satellite. Satellites around other planets are also referred to as moons.

Nucleus   The solid body of a comet. Varying in composition, but always containing low-temperature volatiles, cometary nuclei can measure anything from a few hundredths of meter to a few hundred kilometers in width.

Oort Cloud   A hypothesized shell of countless billions of mainly icy planetesimals (cometary nuclei) that surrounds the Solar System at a distance of around 50,000 au (almost a light-year away). It is the source of many new comets. Named after the astronomer Jan Oort who postulated its existence in the mid-twentieth century.

Opposition   Viewed from Earth, the position of a Solar System object when its celestial longitude is 180° to that of the Sun (i.e., opposite to the Sun).

Orbit   The gravitationally induced path of one object around another.

Parallax   The change in an object's apparent position with respect to more distant objects, caused by a change in viewing angle. Parallax affects the apparent position of an asteroid or comet when viewed from widely separated points.

Perigee   The point in an object's orbit when it is nearest to Earth.

Perihelion   The point in an object's orbit when it is nearest to the Sun.

Perturbations   The gravitational influence of one celestial object on another. Comets in our Solar System are prone to perturbation by the gravity of the planets, particularly Jupiter, which alters their orbital paths. (Enough is known about perturbation that it can be factored in to calculations about objects' future paths.)

Planet   A large non-stellar object in orbit around a star. Our Sun has eight planets, five dwarf planets and many thousands of minor planets.

Reflector   A telescope that uses a large primary mirror to collect and focus light.

Refractor   A telescope that uses a large lens to collect and focus light.

Retrograde motion   The outer planets exhibit retrograde motion as seen from Earth when they move in the sky from east to west rather than the normal west to east motion. This is due to Earth, which has a faster speed than the outer planets, catching up with the outer planet and then pulling away as it orbits the Sun. Comets can also appear to undergo retrograde motion.

Right ascension (RA)   A position on the celestial sphere equivalent to terrestrial longitude, measured eastward from the celestial meridian in hours, minutes and seconds. The celestial meridian intersects the celestial equator at the vernal equinox, the point on the celestial equator in Pisces where the Sun crosses from north to south. There are 24 h of RA, and the 24-h line is always taken as 0 h.

Satellite   Any object in orbit around a larger body. Most planets have natural satellites.

Seeing   A measure of the quality and steadiness of an image seen through the telescope eyepiece. Seeing is affected by atmospheric turbulence, caused largely by thermal effects.

Solar System   The Sun and everything that orbits the Sun, including the planets and their satellites, asteroids and comets.

Spectroscope   An instrument that splits up electromagnetic radiation into its different wavelengths and is used to analyze how the intensity of radiation varies with wavelength. Different elements and chemical compounds can be identified spectrographically.

Spectrum   The splitting of electromagnetic waves into its various component wavelengths, as light is split into the colors of the rainbow, is called a spectrum. A spectrum can be analyzed to reveal many properties, including surface temperature, radial velocity, chemical composition, and magnetic field strength.

Star   A huge ball of incandescent gas shining by nuclear fusion. Our Sun is a star.

Sublimation   When a solid volatile at low temperature in space is heated sufficiently by the Sun it turns into a gas, avoiding the liquid state. The process is called sublimation.

Sun Our nearest star.

Telescope An instrument that collects and focuses electromagnetic radiation— from long wavelength radio waves, through visible light to short-wavelength gamma rays. Optical telescopes produce magnified images of distant objects by using lenses and/or mirrors to collect and focus light.

Universal Time (UT) The standard measurement of time used by astronomers all over the world. UT is the same as Greenwich Mean Time, and it differs from local time according to the observer's position on Earth and the time conventions adopted in that country.

Zenith The point in the sky directly above the observer. It has an altitude of 90°.

About the Author

Peter Grego began his astronomical writing career by producing around 50 articles on astronomy and observing for the Birmingham (UK) Astronomical Society Journal (1982–1991). Through the years he has had dozens of articles appear in various publications, including *Popular Astronomy*, *Sky & Telescope*, *Astronomy Now*, *Sky at Night* magazine, *Amateur Astronomy and Space Sciences*, *Fortean Times*, *Selenology*, *Gnomon*, *The Lunar Observer*, *The New Moon*, *The Encyclopedia of Astronomy and Astrophysics*, *Window on the Universe* CD-ROMs, and *Hoshinavi*. He has contributed to several books by Ian Ridpath, Brian Jones, and Robin Scagell and revised the lunar material for both the latest edition of *Norton's Star Atlas* and for the two latest editions of Philip's *Universal Atlas*.

He currently edits the following astronomy publications: *Luna*, journal of the SPA Lunar Section (since 1984); *Popular Astronomy* magazine (since 2000); and the BAA Lunar Section Circulars and the Bulletin of the Society for the History of Astronomy.

Peter also finds time to write and illustrate the monthly "MoonWatch" page for *Astronomy Now* magazine (since 1997). He has given many talks to astronomical societies around the UK and has been featured on a number of radio and television broadcasts.

Finally, Grego is the author of numerous astronomy books, including *Collision: Earth!* (Cassell, 1998), *Moon Observer's Guide* (Philips/Firefly, 2004), *The Moon and How to Observe It* (Springer, 2005), *Need to Know? Stargazing* (Collins, 2005), *Need to Know? Universe* (Collins, 2006), *Solar System Observer's Guide* (Philips/Firefly, 2005), *Exploring the Earth/Exploring the Moon/ Discovering the Solar System/ Voyage Through Space/ Discovering the Universe* (five books in the

P. Grego, *Blazing a Ghostly Trail: ISON and Great Comets of the Past and Future*, 285
The Patrick Moore Practical Astronomy Series, DOI 10.1007/978-3-319-01775-4,
© Springer International Publishing Switzerland 2014

QED Space Guides series, 2007), *Venus and Mercury and How to Observe Them* (Springer, 2008), *The Great Big Book of Space* (QED 2010), *Galileo and 400 Years of Telescopic Astronomy* (Springer, 2010), and *Mars and How to Observe It* (Springer, 2012). *Cornwall's Strangest Tales* (Portico, 2013) and *My Little Book of Space* (QED, 2013).

Index

P. Grego, *Blazing a Ghostly Trail: ISON and Great Comets of the Past and Future*,
The Patrick Moore Practical Astronomy Series, DOI 10.1007/978-3-319-01775-4,
© Springer International Publishing Switzerland 2014